RICH HARVEST

RICH HARVEST:

A History of the Grange,

1867–1900

D. SVEN NORDIN

UNIVERSITY PRESS OF MISSISSIPPI

JACKSON

THIS VOLUME IS AUTHORIZED
AND SPONSORED BY
MISSISSIPPI STATE UNIVERSITY

CONTENTS

The New York *Sun* once described a granger as "a person little known but much discussed. He is very myste[ri]ous." Although this description was given in 1874, it is appropriate a century later. The term *granger* has come to have a dual meaning. To individuals affiliated with the Patrons of Husbandry, the word connotes membership in their brotherhood; to students of American history, it has a much broader application. Historians using the term refer not so much to members of a particular agricultural organization as to farmers whose desires for change were so great during the 1870s that they constituted a movement of protest against alleged oppressors.

Moreover, scholars generally use the phrase Granger Movement to refer to the agrarian discontent of the seventies and to the corresponding actions taken by farmers to rid themselves of their nemeses. According to the interpretation that accompanies this usage, abusive railroad practices, corrupt politicians, and vulturous businessmen are singled out as the Grange's main concerns; granger laws are still considered to be the results of the order's protests; and Solon J. Buck's study of grangerism written sixty years ago is still accepted as a creditable account of the movement.

The years between the organization of the Grange in 1867 and the society's decline in the Middle West and South around 1880 have traditionally been referred to as the period of the Granger Movement. Writers of textbooks, monographs, and scholarly articles have accepted without challenge Buck's time guidelines as laid down in his "classic" study, *The Granger Movement,* even though membership records and organization activities clearly refute his thesis.

Although the traditional interpretation still prevails, a number of attacks have been made upon it. Professors Benson, Kolko, Nash, and Miller have shown the relative insignificance of the Patrons of Husbandry, or the Grange as it is more commonly called, in the struggle to regulate rail carriers. No one, however, has come forward to reexamine the brotherhood per se and its activities, place them in proper perspective, and give them continuity.

The purposes of this study, therefore, are fourfold. It relates the history of the Patrons of Husbandry in the nineteenth century, shows how the Grange and its growth patterns during the period reflected the influence of one area of settlement upon others, illustrates that there were really two granger movements, and gives credence to the statements made repeatedly by grange leaders that their order was primarily a social and educational fraternity for farmers and their families rather than a medium for political and economic activities.

One theme runs through the entire study. The objectives established by the order's founding fathers remained the bases of the Grange's activities. There were deviations to be sure, but they were overshadowed by the conformity to the brotherhood's statement of purpose. In other words, grangers were to benefit from interest in political, social, economic, and educational matters; but political and religious partisanship were not to be parts of their normal routine.

I wish to express my appreciation to the people who have aided me in preparing this study. In particular, I am most grateful to Dr. Roy V. Scott, my major professor at Mississippi State University, for his beneficial suggestions, counsel, supervision, and patience. I also want to thank Dr. Glover Moore for carefully reading the manuscript. I equally appreciate the valuable assistance given by Professors Gerald Prescott, Howard Schonberger, Stuart Noblin, and William Barns and by Mr. Douglas Bakken. For opening the inner gate to me, the members and officers of Wilmington Grange No. 1918 of Harveyville, Kansas, are worthy of special notation. Their bending of rules to permit me to observe the secret ritual gave me a better understanding of grange ceremony and procedure.

The librarians and staff at the Wisconsin State Historical Society Library deserve special praise for their understanding. Although their names evade me, acknowledgement must also be made to the countless individuals who helped me at the following places: University of Arkansas, University of Mississippi, Mississippi State University, Mississippi State Department of Archives and History, University of Alabama, University of Missouri, Missouri State Historical Society, Nebraska State Historical Society, Iowa State Historical Society, Minnesota State Historical Society, Colorado State Historical Society, University of Wisconsin, Indiana State Historical Society, Indiana State Library, Indiana University, Ohio State University, Ohio State

Historical Society, University of Kentucky, Tennessee State Library, West Virginia University, University of Virginia, United States Department of Agriculture Library, Library of Congress, National Grange Office, University of Maryland, Duke University, University of North Carolina, University of Georgia, University of Illinois, Illinois State Historical Society, John Crerar Library, Newberry Library, Chicago Historical Society, Illinois State Archives, New Hampshire State Historical Society, Clemson University, Cornell University, and the South Caroliniana Society.

Bryant College   D. SVEN NORDIN

RICH HARVEST

CHAPTER I

Like a Mighty River

After the Civil War American farmers and their families were torn by two contradictory forces—Jeffersonian agrarianism and a new industrial urbanism. The monotony and drudgery of rural life served to sharpen the contrast between the two forces and did much to create bitterness among farmers. Awareness of a conflict of cultures had existed before 1861, but the war contributed dramatically to a deepening of the clash. The military experiences of men from rural backgrounds increased their desires for a better life than the farm offered. Northern farmers who remained at home during the war were also affected. Their farms literally became commissaries, and demand for their products rose to new heights. Farmers expanded production, invested in land and machinery at unprecedented levels, and enjoyed prosperity and material comforts that had been hitherto unknown. When peace brought some retrenchment, such farmers were certain to be numbered among the discontented for they were now commercial producers with all the complexities this represented and implied. No longer could they count themselves independent and free. For now, they were slaves to a commercial economy.[1]

Conditions in the South conflicted sharply with those in the North in the postwar period, but the reaction of ordinary farmers would be little different. Depravity and destruction existed in Alabama and Mississippi on a level unknown in Illinois and Indiana, but the former Confederate felt the same forces that disturbed northern agriculturists. Consequently, when northern and southern soldier-farmers "beat their swords into plowshares," they were unprepared to accept the secondary place in American life that the times seemed to assign them. Northern farmers were being forced to adapt to commercial agriculture, with all its implications and ramifications, while southern farmers were attempting to restore commercial farming. In both

[1] Anne Mayhew, "A Reappraisal of the Causes of Farm Protest in the United States, 1870–1900," *Journal of Economic History*, XXXII (June, 1972), 469–75.

instances, farmers were faced with unfamiliar economic pressures. Concerns after the war were railroad tolls, bank rates of interest, implement costs, grain-storage fees, and commodity prices. Put simply, the farmer found himself in a new, cold world, and he was not gladdened by the prospects facing him.[2]

No one understood what had taken place better than Oliver Hudson Kelley, an employee of the United States Department of Agriculture. He had long recognized a void in the lives of the nation's rural inhabitants, and he gained deeper insight by traveling through the South and by talking with southern farmers. He had received a commission to survey the war-torn region in 1866 as a result of a suggestion by Commissioner of Agriculture Isaac Newton to President Andrew Johnson. Wandering along the azalea- and palmetto-lined roads and conversing with natives impressed Kelley in two ways. He noticed how his membership in the fraternal order of Masons enabled him to win the confidence of his hosts, and he was struck by the lack of communication existing in the rural districts.

According to Kelley, he put the two factors together while daydreaming along the banks of the Mississippi River. Thinking about the long stream and its many tributaries brought a picture to his mind. Eight years later he recalled having seen a relationship between the long waterway and the plight of Dixie farmers. His visualization was of a national agricultural society binding scattered members together as the Mississippi is fed by smaller streams.

It took an exceptional person to see similarities between a mighty river and the needs of American farmers, and Kelley was such an individual. He possessed the visionary powers of a prophet, the zeal of a reformer, and the dedication of a monastic. He was also a very restless person. Unlike many crusaders, Kelley was not blinded by his mission. He actively sought advice and willingly accepted compromise. In many respects his life is a study of pragmatism. In other ways he was like his German contemporary, Karl Marx, in that both men shunned their family responsibilities in order to propagate their ideas and to proselyte the masses. Kelley's wife, Temperance, frequently assumed the role of a widow because her husband regularly left her for long periods of time.[3]

2 *Ibid.*, for an excellent description of Civil War agriculture, see Paul W. Gates, *Agriculture and the Civil War* (New York, 1965).

3 Oliver H. Kelley, *Origin and Progress of the Order of the Patrons of Husbandry in the United States: A History from 1866 to 1873* (Philadelphia, 1875), 1–56.

Kelley had always been a rambler. He was born January 7, 1826, in Boston, Massachusetts. After attending public schools and Chauncey Hall Academy, he sought greener pastures, going first to Chicago. He worked there as a drugstore clerk and as a reporter for the Chicago *Tribune*. But, he did not long remain in the bustling Lake Michigan city. He pushed on to Peoria, where he learned the telegrapher trade. After working a short time in central Illinois, the young drifter again traveled, this time crossing the Mississippi River and going to Bloomington, Iowa. Life there did little more to satisfy the drifting Kelley than that of previous stops. Therefore when tales from Minnesota of pioneer farming opportunities reached him, the adventurous young man decided agriculture was for him. So moved, Kelley went to Minnesota and selected land near Itasca. Life here must have been agreeable for the handsome young easterner because it did not take him long to adjust well to frontier life and farming. In the process, Kelley married and with his wife managed a productive farm. Among his neighbors, Kelley was quickly gaining recognition for his skill and scholarly knowledge of scientific agriculture. In 1853 he and his Itasca admirers organized the Benton Agricultural Society; Kelley served as secretary of the group.

As Kelley's knowledge increased, so did his reputation. In 1864 he was offered a position with the fledgling United States Department of Agriculture, and he accepted the challenge. This appointment proved to be fateful because it resulted in his selection as the presidential inspector of postwar southern agricultural conditions.[4]

While still on his fact-finding mission for the chief executive, Kelley sought to ascertain what reaction his plan for a farm order would receive. P. H. Woodward of Savannah, Georgia, was the first man to learn of his scheme, and the response must have been encouraging for, upon returning to Washington, Kelley immediately discussed his idea with Dr. John Trimble, an old friend, and he also mentioned it in a letter to Carrie Hall, a favorite niece. In her reply Miss Hall offered her uncle some valuable advice. She suggested that he give women full membership privileges in the proposed society.

Kelley then returned to his Minnesota homestead in mid-1866, but his stay was short. He found tilling fields and tending livestock dull and routine, and he longed for an opportunity to leave Minnesota. Moreover, the rigorous demands of farming did not give him enough

4 Lida S. Ives (comp.), Data Relating to the Patrons of Husbandry, Patrons of Husbandry Collection, Minnesota Historical Society, St. Paul.

time and energy to develop his plan. Fortunately, he did not have a very long wait; in the fall of 1866 a vacancy developed in the Post Office Department, and it was offered to Kelley. He promptly accepted the appointment and returned to Washington.

Working in the nation's capital brought Kelley into contact with many influential men. Their assistance and significance in the order's history has been largely ignored. Kelley's role as the "father of the Grange" has been exaggerated because, without the aid of associates, his dream might not have been materialized. In reality he had been unable to formulate satisfactory plans for an agricultural fraternity by himself; the task was too formidable for him. He wisely recognized his limitations and turned to friends for assistance. They each contributed in some way to the general refinement of Kelley's initial plans.

William Saunders, chief pomologist and horticulturist in the Department of Agriculture, was especially instrumental. His letter book clearly shows that he worked closely with Kelley. Saunders' contributions were two-fold. He drafted a general organizational plan calling for local, county, state, and national bodies, which became, with only minor alterations, the basis of the society's system of chapter stratification. His presence was also strongly felt in one other way. Saunders' reputation was such that he was well known among the nation's leading gardeners and fruit growers, and he corresponded with many of them, including Anson Bartlett of North Madison, Ohio, William Muir of St. Louis, and A. S. Moss of Fredonia, New York. By knowing these men and others, Saunders was in a position to advise Kelley to present his plans to them for their advice. The resulting exchanges between Kelley and Saunders' colleagues proved very profitable; Moss and Bartlett later helped Kelley, too, when he came to their states to organize grange chapters. They introduced him to local agricultural leaders and gave him free lodging. For his efforts Saunders, and not Kelley, was singled out by some contemporaries as the one man most responsible for the society's creation.[5]

Some of Kelley's Washington associates assisted in other significant

[5] W. L. Robinson, *The Grange, 1867–1967; First Century of Service and Evolution* (Washington, 1966), 1–4; William Saunders Letter Book, National Grange Library, Washington, D.C.; Conversation with Harry Graham, Washington, D.C., December 2, 1966; Madison *Western Farmer*, December 20, 1873; Denver *Rocky Mountain Weekly News*, March 4, 1874; William D. Barns, "Oliver Hudson Kelley and the Genesis of the Grange: A Reappraisal," *Agricultural History*, XLI (July, 1967), 229–242.

ways in the formation of the order. When the Minnesotan encountered difficulty with ritualism, he consulted John R. Thompson, a high-degree Mason. Thompson examined and revised the original rites and then drafted additional ones for the sixth and seventh degrees.

Additional refinements were supplied by others in Kelley's Washington circle of friends. William M. Ireland framed the society's constitution and bylaws, prepared the order's proceedings and pamphlets for printing, and checked galley proof. The Reverend Aaron B. Grosh added a spiritual and musical touch to grange ritualism by writing prayers and collecting songs for use in meetings. John Trimble was the critic of the group. He was probably the least appreciated, but in some ways the most helpful, of Kelley's companions. Trimble was dubbed "wet blanket" by his partners because he unhesitantly sneered at their work if he felt it was justified. His caustic remarks were nevertheless invaluable as they made the members of Kelley's group more cognizant of weakness and because they resulted in the correction of innumerable flaws.

Partial credit for organizing the order should also be given to Francis M. McDowell. He was the financial wizard of the group. His wife's appraisal of the Grange's creation gives a keener insight into the role each man played than any simple statement attributing development solely to Kelley's efforts: "To sum up, we have here a propagandist [Kelley], an organizer [Saunders], a ritualist [Thompson], a parliamentarian and journalist [Ireland], a man of God [Grosh], a critic [Trimble], and a financier [McDowell]." [6]

Still, caution and restraint are necessary in assessing the inventive genius of these seven men; quite often their new ideas were merely old ideas reshuffled. The development of agricultural societies in the United States predates the births of the grange founders. According to the reports of the commissioner of agriculture, over nine hundred groups flourished in 1867. As noted, Kelley himself had affiliated with a local Minnesota farmers' club in 1853. One might be tempted to evaluate the significance of these organizations upon grange development, but this is impossible because few manuscripts linking possible relationships exist. But it is safe to assume that the Grange was more a result of amalgamating characteristics of existing societies and Ma-

[6] Robinson, *The Grange, 1867–1967*, 3–4; Kelley, *Origin and Process of the Order*, 1–56.

sonry than a case of ingenuity. In other words the seven founding fathers were experts in the art of eclecticism.[7]

A final step in the founders' work was their naming of the secret brotherhood—the Order of Patrons of Husbandry. Since individual chapters, whether local, county, state, or national, were known as granges, the society has often been referred to as the Grange. The official title of the organization and its common name have been interchanged so many times that either name may be used without confusion or offense to members.

The plan for the Grange was really quite simple. Each level of the four-plane grange structure had thirteen officers—master, overseer, lecturer, steward, assistant steward, lady assistant steward, chaplain, treasurer, secretary, gatekeeper, ceres, pomona, and flora. Their duties were not explicitly outlined by the founding fathers in the grange constitution. Although the main obligation of each official was to carry out the laws of the society, specific responsibilities were delegated to each office and the determination of these duties rested with each body. They therefore varied somewhat from grange to grange. All officers were elected by ballot. Local or subordinate granges annually selected their leaders, state grange officers held their posts for two years, and national granges chose new functionaries every three years.[8]

With allowance for some minor differences, duties of Texas State Grange officers as prescribed by the state constitution resembled those adopted elsewhere by other bodies. In Texas, as in other states, the master was presiding officer at all state meetings, and it was his duty to fill all vacancies by appointing replacements. The overseer was next in importance, and he took the reins of power whenever the master was not present. A person designated lecturer was the order's program chairman. He planned and led the part of the grange program devoted to education, and should his superiors be absent, he was instructed by the constitution to direct sessions. Following the lecturer in line of succession was the steward. His usual duties included making "all necessary arrangements for the comfort and convenience of the members" and caring for the regalia and furniture of the grange. In his custody were symbolic tools used in meetings, flags, stands and

[7] U.S. Commissioner of Agriculture, *Report*, 1867 (Washington, 1868), 364–403; Frederick Merk, "Eastern Antecedents of the Grangers," *Agricultural History*, XXIII (January, 1949), 1–8; J. A. Cramer, *The Patron's Pocket Companion in Four Parts* (Cincinnati, 1875), 5.

[8] Cramer, *The Patron's Pocket Companion*, 17, 24.

podiums, song books, and sashes worn by officers. As their titles imply, the assistant and lady assistant stewards helped the steward with his work. A chaplain gave all prayers and read Scripture at sessions. According to the same Texas constitution, business and financial records were kept by the secretary and the treasurer, with both officers having to adhere to certain standards established by the rules of the Lone Star society. To insure proper conduct of these offices, the treasurer had to post a personal bond "in a sufficient amount to secure the money" in the treasury, and the secretary was required to file a quarterly report with the master. This statement was to be a statistical account of "the information in his office." Finally, a gatekeeper guarded the "outer gate" and requested the secret password from all would-be participants before allowing their entry into closed meetings. Exact duties for the other three officers were not defined.[9]

Looking at Grange ritual, we see reflected not only Kelley's first-hand knowledge of farming but also his interpretation of the primary needs of rural Americans. He had vivid memories of the grueling work involved in breaking virgin soil and of the endless cycle of a farmer's daily chores. Machines like the reaper and thresher had alleviated some of the punishing toil and permitted farmers to increase their acreage and output. At the same time technological progress forced farmers to compete in more price-conscious markets. As a result farmers still worked long hours after the Civil War and enjoyed few diversions. Religious services and visits to town offered the most consistent escapes from daily routine. Additional outlets for social intercourse existed, but they often were ineffective because they lacked binding force to draw agricultural families together. To Kelley this need was the primary weakness of farmers' clubs. "Country and town societies and clubs are interesting for a while, but soon lose their interest, and I see nothing that will be lasting, unless it combines with it the advantages which an Order similar to our Masonic Fraternity will provide." [10] Kelley was pinning Grange success to the influence of oaths of loyalty and brotherhood along with the spell cast by mystical and secret ritual. These intangible, almost metaphysical forces would, if applied to rural associations, arouse American farmers from their lethargy and propel them into a euphoric state of excitement and imagination. At least these were the dreams of Oliver Hudson Kelley. Con-

9 Texas State Grange, *Constitution* (Waco, 1874), 4–9; for other constitutions consulted, see those listed in the bibliography.
10 Kelley, *Origin and Progress of the Order*, 22.

sequently, Grange plans calling for elaborate rites similar to those in other fraternal orders were an attempt to reduce sterility and colorlessness associated with traditional farmers' organizations.

Grange ritual inspired as it was by Greek and Roman mythology and by biblical lessons included seven degrees: Faith, Hope, Charity, Fidelity, Pomona, Flora, and Ceres. Each degree had its own distinct ceremony of installation, and responsibilities within the order increased with each elevation on the grange ritual ladder. For example, all business conducted by local chapters was enacted by fourth-degree members. Each of the first four degrees represented a different season of the year, and all four were conferred by subordinate granges. Corresponding male and female titles of address accompanied each lower degree: Faith—laborer and maid; Hope—cultivator and shepherdess; Charity—harvester and gleaner; Fidelity—husbandman and matron. As originally planned, the fifth degree was to be conferred by both state and county granges. When the former body bestowed the honor, only local masters and fourth-degree wives of masters were eligible for Pomona status. But when county and district granges granted the degree, the number of eligible persons was expanded to include past masters and a maximum of three fourth-degree members from each local chapter. To be a Flora granger, one had to be either a state master, a fifth-degree wife of a state master, or a member of the National Grange executive committee. Ceres signified the highest honor given by the Patrons of Husbandry; only sixth-degree members who had served the National Grange for one year were to receive promotion to this select circle. Inclusion in this body meant added responsibilities and privileges. Those in the Ceres elite were entitled to govern secret activities of the order, to hear evidence, and to render decisions in impeachment cases against wayward National Grange officers.[11]

Grange ritual was not limited to granting of degrees; it entered into almost every aspect of the order's activities. Even the seemingly simple act of entering a grange hall was covered. To gain admission, members rapped a predetermined number of times on the outer gate and then replied correctly to the gatekeeper's questions. Adhering to a code permitted members to come into the anteroom. In order to enter the main assembly room, grangers once again knocked an appropriate number of times. The number of strokes upon the inner door corresponded with the degree under which the meeting was being held.

11 Cramer, *The Patron's Pocket Companion*, 14–16; Robinson, *The Grange, 1867–1967*, 44–47.

Thus, to gain admission to a session conducted in the third degree required three raps in correct cadence. The assistant steward guarded the inner door and judged the incoming member's adherence to entrance specifications. Satisfactorily complying with ritual demands permitted the entrant to present himself to the overseer. Upon giving correct salute to that officer, the patron was granted permission by an exchange salutation to participate in the meeting. But he could not take any seat that suited his fancy. According to grange ritual, every officer had a designated post to man, and all others were required to sit outside the perimeter occupied by elected officials. Note the officer placement in a grange meeting on the following page.[12]

The final preliminary preparations were completed by the organization of Potomac Grange No. 1 on January 8, 1868. Previously, Kelley and his colleagues had met informally to draft plans, but now they were ready to test their scheme in a laboratory grange. Their first official meeting was in a crowded room on Ninth Street in the nation's capital. There, however, organizers found the chamber too congested to test their ritual adequately, so they removed to the Union League Hall for their next session. Convening regularly from January through March of 1868, as the seven fathers did, they were able with assistance provided by impartial observers from the Post Office Department, to rehearse the ritual and to determine what flaws had to be corrected. In these trial runs, a number of rough edges were smoothed out, and the organization and its ritual were generally perfected. Thus, by the end of March, Kelley, the self-chosen missionary of the group, was impatiently awaiting the opportunity to spread the grange gospel to the nation's farmers.[13]

---

[12] Patrons of Husbandry, Manuscript headed "To enter a grange hall," University of Virginia Library; Cramer, *The Patron's Pocket Companion*, ii. See diagram on page 12 for officer placement in a grange meeting.
[13] Kelley, *Origin and Progress of the Order*, 61–90.

PLAN OF A GRANGE ROOM *

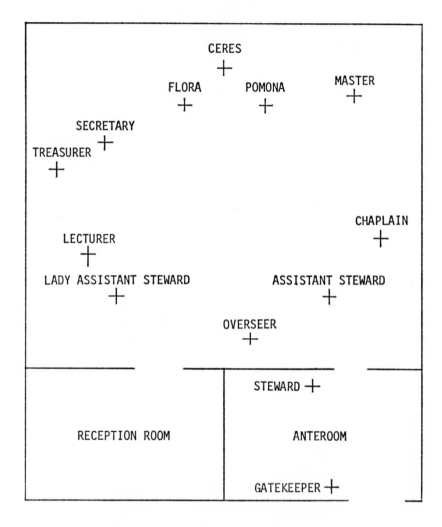

CHAPTER II

Organizational Developments

1868–1900

By the spring 1868 the Patrons of Husbandry had been launched; presumably, it lacked only members. Kelley himself assumed the responsibility for building the order from the grassroots. There followed a swing through the East and Middle West, but results were less than satisfactory. Changes in objectives had to come before the organization could go into its well-known, militant stage, which has generally been known as the Granger Movement. Many historians have assumed that when that phenomenon faded from the scene, grangerism was dead. Admittedly, decay did set in, but subsequently a second Granger Movement appeared, which in the long run quite probably was more significant than the first one.

The years between the organization of the Grange in 1867 and the society's decline in the Middle West and South around 1880 have traditionally been referred to as the period of the Granger Movement. Most writers of textbooks, monographs, and scholarly articles have accepted without challenge Solon J. Buck's time guidelines as laid down in his "classic" study, *The Granger Movement,* even though membership records and organization activities clearly refute his thesis. Why Buck's treatment of the Patrons of Husbandry has met the test of time without serious refutation is an enigma. Certainly, other books covering different aspects of the late nineteenth-century agrarian revolution have not escaped critical evaluation and revision. John D. Hicks's *The Populist Revolt* has repeatedly been dusted off for historical analysis. Like soldiers going to war, historians have battled doggedly over the fine points of Hicks's study and have divided into rival camps over interpretation.[1]

---

[1] Solon J. Buck, *The Granger Movement: A Study of Agricultural Organization and Its Political, Economic, and Social Manifestations, 1870–1880* (Cambridge, 1913); John D. Hicks, *The Populist Revolt* (Minneapolis, 1931). For conflicting interpretations and defenses of this work, see Richard Hofstadter, *The Age of Reform, From Bryan to F. D. R.* (New York, 1955); Norman Pollack, "Hofstadter on Populism: A Critique of *The Age of Reform,*" *Journal of Southern History,*

An understanding of the Granger Movement and of its continuing significance requires a general knowledge of conditions affecting farmers throughout the nation during the post-Civil War years as well as an awareness of the frontier concept. By definition and application, Frederick Jackson Turner used the word *frontier* as "the outer edge of the wave—the meeting point between savagery and civilization." He treated the effects of new environments upon pioneers' lives and general American development, but he largely ignored what happened in the process to old frontiers and their populations when they were faced with competition from newer areas of settlement. Consequently, he did not discover innovative changes which had been forced on older sections to make them competitive with virgin territories.[2] The Grange and its growth patterns in the nineteenth century reflected the influence of one area of settlement upon others. As farmers settled the Great Plains, they had to adjust to new conditions. Corresponding adjustments also were made east of the Mississippi River. By the 1880s, easterners had to revise their agricultural practices to survive the onslaught from prairie competition. To combat the advantages enjoyed by settlers of the Plains, easterners turned to the Patrons of Husbandry, seeking legislative and other assistance.[3]

Since Buck failed to grasp the correlation between eastern and western agriculture and since he ignored the causal relationship between the two sections, he did not discover that there were really two granger movements. One covered the years 1870–1880 and primarily affected the West and South; the second spanned the years 1880–1900 and displayed its greatest vigor in New England, New York, Pennsylvania, and the eastern Middle West. By dealing only with the former move-

---

XXVI (November, 1960), 478–500; Norman Pollack, "The Myth of Populist Anti-Semitism," *American Historical Review*, LXVIII (October, 1962), 76–80; Norman Pollack, *The Populist Response to Industrial America: Midwestern Populist Thought* (Cambridge, 1962); Walter T. K. Nugent, *The Tolerant Populists; Kansas, Populism and Nativism* (Chicago, 1963); Theodore Saloutos, "The Professors and the Populists," *Agricultural History*, XL (October, 1966), 235–54. For proof of the unadvanced state of granger historiography, check the treatment afforded the order and the appraisal given Buck in John A. Garraty, *The New Commonwealth, 1877–1890* (New York, 1968), 52–53, 63, 66, 69, 71, 75–76, 114, 163, 339.

2 Frederick J. Turner, *Frontier and Section* (Englewood Cliffs, N.J., 1961), 37–62.

3 A good description of what happened in one state is offered by Paul W. Gates in "Agricultural Change in New York State, 1850–1890," *New York History*, L (April, 1969), 127–40. See also Rexford Sherman, "The New Hampshire Grange 1873–1883," *Historical New Hampshire*, XXVI (Spring, 1971), 2–25. If nothing else, Sherman shows similarity of granger concerns in New Hampshire with those elsewhere.

ment, Buck did not show that the eastern resurgency saved the organization from expiring after its decline in the West and South. The dual impact of the frontier upon late nineteenth-century agricultural history offers a partial explanation for the inadequacy of Buck's treatment of the Granger Movement.[4]

The post-Civil War years were marked by unprecedented agricultural expansion. Farmers staked off more new homesteads and grew more food and fiber than in any previous period in the nation's history. The amount of farmland increased from 407,213,000 acres in 1860 to 838,592,000 in 1900, while the number of farms rose from 2,000,000 to 5,737,000.[5] Plainly, men of all ages had followed Horace Greeley's famous advice. Most of the newly plowed soil was located west of the Mississippi River. Census figures illustrate the westward movement of agricultural wealth.

Table 1

VALUE OF ALL FARM PROPERTY
(in Billions of Dollars) [6]

Section	1870	1880	1890	1900
North Atlantic	3.684	3.197	2.970	2.951
South Atlantic	.926	1.053	1.333	1.454
North-Central	5.136	6.108	8.518	11.505
South-Central	1.134	1.290	1.891	2.816
Western	.245	.532	1.371	1.715
United States	11.125	12.181	16.082	20.514

The sharp rises in the north-central and western states and the steady decline in the north Atlantic area show that agricultural expansion took place in the West and that contraction and stagnation transpired in the East.

Expanded acreage during the postwar period was more a bane than a blessing for American farmers. The results were overproduction and lower prices. Market growth from 1865 to 1900 simply did not match

---

[4] For treatments of the frontier and its impact upon late nineteenth-century agricultural history, see Gilbert C. Fite, *The Farmers' Frontier, 1865–1900* (New York, 1966); Fred A. Shannon, *The Farmer's Last Frontier: Agriculture, 1860–1897* (New York, 1945).

[5] Shannon, *Farmer's Last Frontier*, 51.

[6] U.S. Census Office, *Twelfth Census, 1900, Agriculture* (Washington, 1902), Pt. 1, p. 694.

the increased productivity, and prices spiraled downward in an erratic pattern. Statistics for corn, wheat, and cotton illustrate well how supply and demand adversely affected the nation's crop producers.

Of course, overproduction and falling prices were not the only problems confronting late nineteenth-century farmers. Various other calamities struck agriculturists throughout the United States and left them facing starvation and economic ruin. Those who had settled the "Great American Desert" waged a constant battle against the devastating combination of insects and weather. Ravaging grasshoppers wrought havoc. As settlers watched helplessly, swarming hordes of locusts swept northward like giant black funnel clouds and, with methodical speed and precision, devoured everything edible in their paths. The hungry creatures cleared fields and trees of foliage and ate curtains and bed linens. One farmer in Missouri observed that the winged beasts in his community "have about destroyed every thing

Table 2

CORN [7]

Year	Acreage (m. acres)	Production (m. bu.)	Average Price (cts. per bu. Dec. 1)	Exports (m. bu.)
1867	32.5	768	57	012.4
1869	37.1	874	60	002.1
1871	34.1	992	43	035.7
1873	39.2	932	44	036.0
1875	44.8	1,321	37	050.9
1877	50.4	1,343	35	087.2
1879	53.1	1,548	38	099.6
1881	64.3	1,195	64	044.3
1883	68.3	1,551	42	046.3
1885	73.1	1,936	33	064.8
1887	72.4	1,456	44	025.4
1889	78.3	2,113	28	103.4
1891	76.2	2,060	41	076.6
1893	72.0	1,619	37	066.5
1895	82.1	2,151	25	101.1
1897	80.1	1,903	26	212.1
1899	82.1	2,078	30	.....

[7] U.S. Department of Agriculture, *Yearbook*, 1899 (Washington, 1900), 759.

green in the fields and gardens and have eaten up all the grass in the woods, and are going over the fields and woods again. This has been a terrible month for the farmers." Weather conditions also caused untold hardship. Rampaging rivers in the spring, scorching heat and drought in the summer, and blinding blizzards and bitter cold in the winter did their share to make life trying on the Great Plains.[8]

Kelley and his colleagues knew something of the trouble brewing in rural America. But they believed the mounting problems had resulted from inadequate educational opportunities and lack of social intercourse among rural classes. This belief led them to create an organization that was basically fraternal and educational. Little did they realize that farmers would not interpret their plight in these terms. It was not until after Kelley had left on his first organization

Table 3

WHEAT[9]

Year	Acreage (m. acres)	Production (m. bu.)	Average Price (cts. per bu. Dec. 1)	Exports (m. bu.)
1867	18.3	212	145	025.3
1869	19.2	260	77	053.9
1871	19.9	231	115	039.0
1873	22.2	281	107	091.5
1875	26.4	292	90	074.8
1877	26.3	364	106	092.1
1879	32.5	449	111	180.3
1881	37.7	383	119	121.9
1883	36.5	421	91	111.5
1885	34.2	357	77	094.6
1887	37.6	456	68	119.6
1889	38.1	491	70	109.4
1891	39.9	612	84	225.7
1893	34.6	396	54	164.3
1895	34.0	467	51	126.4
1897	39.5	530	81	217.3
1899	44.6	547	58	. . . . .

[8] Shannon, *Farmer's Last Frontier*, 149–53; Fite, *Farmers' Frontier*, 58, 60–63, 115, 126, 166, 170, 200, 206–207; David Rice Atchison, Books, May 30, 1875, Western Historical Manuscripts Collections, University of Missouri Library, Columbia.

[9] *Ibid.*, 760.

trip on April 3, 1868, that he realized that he and the other founders had misjudged the needs and demands of American agriculturists. They wanted an organization that would defend them politically and would protect them economically. They also desired relief from the new commercial uncertainties now connected with farming. All other concerns were secondary to them at the moment.[10]

Before his departure from Washington, Kelley anticipated no difficulties, since he assumed that crowds of eager farmers would flock to him to learn about the Grange. His expectations were so great that he resigned from the Post Office Department and set out on his journey with only limited funds to sustain him. He was confident that the $15 dispensation fee that he was authorized to collect from each grange organized through his efforts would provide adequate funds to meet traveling expenses.

Unfortunately, his trip netted disappointing results. After a month

Table 4

COTTON[11]

Year	Acreage (m. acres)	Production (m. lbs.)	Average Price (cts. per lb. Dec. 1)	Exports (m. lbs.)
1866	06.3	02.1	. . . .	1.3
1868	07.0	02.4	. . . .	1.3
1870	08.7	04.4	12.1	2.9
1872	08.5	03.9	16.5	2.4
1874	11.0	03.8	13.0	2.5
1876	11.7	04.5	09.9	2.9
1878	12.3	05.1	08.2	3.3
1880	15.5	06.6	09.8	4.4
1882	16.8	06.9	09.9	4.6
1884	17.4	05.7	09.2	3.8
1886	18.5	06.5	08.1	4.3
1888	19.1	06.9	08.5	4.8
1890	20.8	08.7	08.6	5.8
1892	18.1	06.7	08.4	4.4
1894	19.5	09.9	04.6	7.0
1896	23.3	08.5	06.6	6.2
1898	25.0	11.2	05.7	7.5

10 Kelley, *Origin and Progress of the Order*, 91–151.
11 *Ibid.*, 764–65.

of travel and work, Kelley's tally sheet showed only two local chapters established. When he reached his Minnesota farm, he was discouraged and indebted but not defeated.

Although this first mission for the Grange did not accomplish all that Kelley had anticipated, his pioneer efforts were not totally devoid of significance. As he traveled about the country, he began learning firsthand what farmers wanted. Wherever he went, he discovered that they desired protection from "patent right swindlers" and from manufacturers of "worthless machines." From their discussions he concluded that the Grange would have to broaden its objectives in order to attract any large number of farmers.[12]

Upon returning to Minnesota in May, 1868, Kelley found himself in a predicament. His farm was heavily mortgaged, and the Grange was not catching on as he had expected. Furthermore, his associates in Washington were losing interest in the order. Kelley recognized that he was at a crucial crossroads, and his problems presented him with no small dilemma. Should he abandon his dream of uniting farmers, or should he continue to sacrifice his time, energy, and his family's welfare for the cause in which he believed?

Kelley remained at his Minnesota farm for several months, and the experience of living again among farmers helped him to realize what needed to be done. His Itasca neighbors told him that the organization's objectives were too idealistic and that they did not get at the heart of agrarian discontent; the order had no hope unless its objectives were expanded to reflect the needs of rural America and unless certain "flowery" passages of ritual and "squints of a mutual admiration society" were eliminated from it. Kelley, in fact, had already noted that when he tried to interest farmers in the Grange their usual response was "what pecuniary benefit are we to gain by supporting the organization." These facts caused Kelley to make a complete examination of the whole Grange program. Quite clearly, the order would prosper only if he managed somehow to tie membership in the society to economic improvement.[13]

Knowing that the Grange would have to base its membership drives on economic terms did not solve all of Kelley's problems. He still had to develop some plan for assuring financial returns to members, and he had to find some way of acquainting farmers with the organization. After wrestling with the first problem for almost two months, Kelley

12 *Ibid.*, 91–99.
13 *Ibid.*, 111–14, 146–47, 160, 176.

hit upon a solution. On July 12, 1868, he wrote his Washington associates, spelling out how patrons could make agriculture a more rewarding profession by combining economically to choke off middlemen. According to Kelley's scheme granges would assume the responsibilities of middlemen, and farmers once again would control their destinies by regulating the market price of their goods. In other words Kelley was proposing that each grange should become a "Board of Trade" in order to maximize farmers' profits and to give members immediate, pecuniary rewards. This cooperative principle became the basis of future grange business endeavors and it answered the farmers' most searching question about the order's benefits.[14]

Even after the Grange had been broadened to coincide with farmers' demands, Kelley still had to find some means of popularizing the order. His first mission taught him that the number of members recruited by circuit-rider tactics did not justify the expense of traveling from farm to farm. Rural people, he discovered, were too suspicious of strangers to pay money for a charter to an unknown club. Consequently, he did not again attempt to organize farmers by direct means without first acquainting them with the order and its objectives and without receiving their assurance that they were interested in forming a grange. To meet these needs, he turned to the press.[15]

Kelley openly solicited support of newspapermen and editors of agricultural journals, a campaign that proved to be very successful. By August 1, 1868, five leading farmers' magazines agreed to print articles about the order. These included the *Prairie Farmer*, the *Farmers' Chronicle*, the *Ohio Farmer*, *Colman's Rural World*, and the *Farmers' Union*. In addition numerous Minnesota daily and weekly papers indicated their willingness to carry grange news and to publicize the society.[16]

Kelley's determination to make the Grange a popular part of rural society gradually began to produce results. A steady increase in the number of local chapters in the late 1860s was a good indication of growing popularity. By the end of the decade forty-nine active subordinate granges were in operation. Forty of these were in Kelley's home state, while the others were scattered through Illinois, Iowa, Pennsylvania, New York, and Ohio. Moreover, by early 1869 the Grange had become so strong in Minnesota that Kelley was able to test the order's

14 *Ibid.*
15 *Ibid.*, 115–30.
16 *Ibid.*;Minneapolis *Farmers' Union*, January, 1869.

plan for state organization. He called together representatives from several local bodies, and they formed the first state grange. Two months later, on April 13, 1869, the first annual session of the National Grange convened in Washington. Thus, by 1870, the order was progressing reasonably well; not only was it growing more powerful in terms of membership, but it was also functioning on three levels.[17]

Kelley was personally responsible for chartering most of the granges organized in the formative period, but as the order grew his role as organizer changed. After having chartered ten of the first eleven locals in Minnesota, Kelley was able to transfer his organizational duties there to a deputy. Colonel D. A. Robertson, a charter member of North Star Grange No. 1 of St. Paul, was the first such disciple. By responding to calls from farmers interested in forming granges and by personal assistance in establishing locals, Robertson set examples which were followed by organizers appointed later.[18]

The work of Robertson and countless other volunteers who served as grange deputies should not be underestimated. These men deserve much credit for taking the Grange off the drawing board and placing it in national prominence. By crisscrossing the United States and eastern Canada during the early 1870s grange recruiters were able to organize subordinate chapters by the hundreds. In Indiana, for example, the organization had seven general and over fifty county deputies at work by 1874. They issued more than three hundred charters between December 24, 1869, and August 3, 1873.[19]

Grange deputies did equally well in other areas, despite the normal deterrents. T. A. Thompson of New Jersey braved a blizzard to keep two appointments with farmers. He reportedly was satisfied by the "attendance and interest expressed"; in any event, he noted to Master Mortimer Whitehead that his visit had been profitable. In 1872 the National Grange sent Eben Thompson to Canada. By the end of the year he had established nine chapters in Quebec and one in Ontario.[20]

Grange deputies assigned to one state regularly answered inquiries from farmers in neighboring states. The first granges established in

17 Kelley, *Origin and Progress of the Order*, 168, 174–78, 219; A. B. Grosh to R. B. Mayes, May '8, 1869, in John Bull Smith Dimitry Papers, Duke University Library, Durham, N C.

18 Minnesota Granges, Papers, 1868–74, Minnesota Historical Society, St. Paul.

19 Indianapolis *Indiana Farmer*, February 14, 1874; Indiana State Grange, Records of Subordinate Granges, 1869–73, Indiana State Library, Indianapolis.

20 Mortimer Whitehead to William Saunders, March 1, 1875, in William Saunders Collection, National Grange Library, Washington, D.C.; Dominion Grange, *History of the Grange in Canada* (Toronto, 1876), 4.

Kansas were organized by William Duane Wilson of Iowa. In the spring of 1872 he arrived in Kansas to present the order to the state's farmers. After laying a foundation from which the organization could prosper, Wilson returned home. But in December, he visited Kansas again to set up a state grange. Similarly, on July 9, 1873, Colonel D. Wyatt Aiken of South Carolina helped a group of North Carolinians form a state society. This chartering of a state body came only thirty-two days after Aiken had established the first subordinate chapter in the state.[21]

Organizational meetings thereafter generally followed the same pattern. Farmers usually took the first step by informing someone in the Grange that they were interested in the order. Pursuant to a request for information, an agent from the organization wrote interested parties, instructing them to gather at a convenient meeting place such as a church, a lodge hall, or a schoolhouse. The deputy then visited the community and presided over the meeting. If enough men and women were present to constitute a quorum, the grange official appointed a temporary chairman and a secretary. He then helped those attending the planning session to draft bylaws and a constitution. After these had been adopted, he supervised the election of officers and told new members of the order's ritual.[22]

In most cases granges started from scratch, but occasionally they developed from independent farmers' clubs. The first subordinate chapter organized in Colorado came into being in early 1873 after the members of the Clear Creek Valley Farmers' Club had voted to disband their loosely constituted group in order to affiliate with the strongly federated Patrons of Husbandry. Their action resulted in the formation of Ceres Grange No. 1. Similar developments occurred at Appleton and Greenville, Wisconsin. In both places granges sprang from existing farmers' clubs, because their members desired the special advantages offered only by a national organization like the Grange.[23]

Inducing farmers to join the Grange was not always so easy, because some practices of the order conflicted sharply with teachings of

21 Kansas State Grange, "The Grange Movement in Kansas," *Report of Historical Committee* (n.p., 1952), 1–4. North Carolina State Grange, *Proceedings*, 1873, pp. 3–5; 1874, p. 7.

22 North Carolina State Grange, *Proceedings*, 1873, pp. 3–5; Church Hill Grange (Ky.) No. 109, Secretary's Minutes, December 2, 1873, in Church Hill Grange Papers, University of Kentucky Library, Lexington; Nashville *Rural Sun*, April 3, 1873.

23 Denver *Rocky Mountain Weekly News*, March 4, 1874; Madison *Western Farmer*, October 11, 1873.

certain churches. Some religious sects questioned the secret ritualism of the society, and they did not hesitate to warn their members to reject the farmers' fraternal organization. Within the Roman Catholic Church bishops vacillated as to whether grange membership violated a papal bull issued in 1739 placing all secret societies on the banned list. Catholic prelates who opposed the Grange claimed the secret tenets of grangerism prevented complete sacramental confessions, and they threatened to excommunicate those who defiantly challenged the edict. On the other hand, some local church leaders relaxed restrictions against the Grange and permitted members to join the order.[24]

Of course the papal order against secret societies placed many Catholic farmers in an awkward position. In many cases they sympathized with the granger cause, but they were unable to show their active support because they did not dare to disregard the authority of the Church. Some attempts, however, were made to circumvent the ban.

The rural Catholics of Olmstead County, Minnesota, were especially enterprising. Calling their fellow churchmen together at Rochester in September, 1873, they discussed their role in the agrarian unrest sweeping the nation and adopted resolutions explaining their estrangement from the Patrons of Husbandry. They emphatically stated that their aloofness must not be interpreted as a stand "in favor of monopolists and against the suffering classes," because they were completely sympathetic with the grangers' fight against "the exorbitant exactions of the railroad and other corporate monopolies, and against class legislation calculated to enrich the few at the expense of the many, and against political corruption." The only reason they did not affiliate with the alliance was that they were Catholics and were therefore committed not to take the society's oath of allegiance and not to participate in secret activities. At this meeting, which was addressed by grange orator Ignatius Donnelly, those present formed the Catholic Farmers' and Laborers' Association of Minnesota—a body which failed to emerge from the embryo stage.[25]

---

24 Fritiof Ander, "The Immigrant Church and the Patrons of Husbandry," *Agricultural History*, VIII (October, 1934), 158.

25 Fergus MacDonald, *The Catholic Church and the Secret Societies in the United States* (New York, 1946), 75–79, 210–11; James J. Pillar, "Catholic Opposition to the Grange in Mississippi," *Journal of Mississippi History*, XXXI (August, 1969), 216–28. Father Pillar's article highlights the division among leading churchmen in the archdiocese of New Orleans and the absence of a definite position by the Holy See in Rome. In the final analysis, at least for Mississippi, Catholic opposition when it came mattered little, for there were too few adherents of the faith for them to have much impact on the Grange's overall development.

Nor was the Roman Catholic Church alone in its fight against the Grange; many Protestant denominations were also aligned against the order. The following table shows those sects in open opposition to the Grange on religious grounds.

No religious body assaulted the Grange more vigorously than did the National Christian Association. This group had been formed in 1868 by concerned Protestant ministers who felt that secret societies were corrupting national moral fiber and were violating New Testament teachings. They assailed all such groups, comparing their activities with those of Satan. Most of their attacks appeared in the association's weekly journal, the *Christian Cynosure,* a publication that was filled with emotional tirades and exposés which explained to God's servants why they should not belong to any secret organizations. Being a closed society with confidential rituals, the Grange naturally was singled out by the editors. They consistently damned the organization because to them it represented "the last hope of the devil." [26]

Contributor H. A. Preus captured the general spirit of the publish-

Table 5

PROTESTANT DENOMINATIONS
ON RECORD AS GRANGE OPPONENTS

Denomination	Number of Clergymen	Number of Members
Reformed Presbyterians	100	9,726
United Presbyterians	595	74,833
Associated Presbyterians	12	1,162
United Brethren	1,886	131,859
Free Methodists	145	6,113
German Baptists	1,048	200,000
Lutherans:		
Norwegian-Danish Conference	48	. . . .
Augustana Synod Swedish	93	30,127
Evangelical Synodical Conference	930	187,873 [27]
Total	4,857	641,693

26 *Ibid.,* 13; Chicago *Christian Cynosure,* April 9, May 14, 28, October 8, 1874; Ander, "The Immigrant Church and the Patrons of Husbandry," 158.
27 National Christian Association, *A Brief History of the National Christian Association* (Chicago, 1874), 28.

ers' opinions of the Patrons of Husbandry when he wrote that he was "fully convinced that no Christian can join or remain a member of this society without making himself guilty of denying Christ and other great sins." His article, like most others appearing in the *Christian Cynosure*, relied upon passages from the Bible to support a contention that the Grange was indeed the work of enemies of God. He cited numerous passages which supported his charge that Christianity and secret societies did not mix.[28]

Although it is difficult to measure the influence of the National Christian Association's crusade upon the nation's farmers, there is, nevertheless, sufficient evidence linking passionate appeals against secret societies with agrarian suspicions of the Grange to draw some conclusions. Certainly, many foreign-born clergymen read the *Christian Cynosure* and heeded its warnings about the Patrons of Husbandry. Their publications contained many articles denouncing the order, and they regularly preached sermons about the inherent evils of grange membership.[29]

The lack of foreign names on grange membership rolls suggests that many immigrants must have accepted their preachers' emotional arguments against the order. Scandinavian- and German-Americans evidently listened to their church leaders' objections to secret societies, because most foreigners from northern Europe refused to join their native-born neighbors in their fight for "justice and improvement." Consequently, because the Grange drew little active support from immigrant-farmers, one may assume with some degree of certainty that the agrarian uprising of the late nineteenth century was basically a nativist revolt and not a general rebellion of rural classes.[30]

28 Chicago *Christian Cynosure*, October 8, 1874, April 9, May 14, 28, 1874.
29 Ander, "The Immigrant Church and the Patrons of Husbandry," 158–67.
30 *Ibid.* The information in this paragraph was also based upon an ethnographical analysis of grange membership records contained in roll books, agricultural-journal notices about newly chartered granges, and state and national proceedings and upon conclusions expressed by Gerald L. Prescott, a University of Wisconsin graduate who did research on grange membership in Wisconsin. Prescott utilized quantification techniques to determine who the average Wisconsin granger for the mid-1870s was. From the computer printout, he found the typical member of the state body to be a middle-aged "WASP" raising primarily spring wheat on a 156-acre farm assessed at approximately $5,000. This individual produced some corn for feed, kept a small livestock herd, and milked cows for home consumption. Produce from the average granger operation was about $1,000. In relation to other farmers in his home county, the state granger was slightly above average. Caution must be exercised before drawing too many generalizations about grangers nationally from Prescott's work because he concerned himself with only one state which might not have been typical.

Of course there were exceptions; some farmers openly defied their churches' edicts against secret societies, while others were more subtle. Kansan James Hanway cautiously used the pseudonym, "Mirage," when he defended his affiliation with the Grange in the letters-to-the-editor column of the Garnett *Weekly Journal*. Hanway, like other members of his church, had been very critical of the Grange, but later he reversed his stand: "Now, with all frankness and honesty . . . I must whisper in the ear of my most loving and affectionate brothers and sisters of our church, that they are only creating a tempest in a tumbler, in their opposition to this new social order, called the Patrons of Husbandry." Hanway's circumspection apparently was justified because seven Iowa farmers had been "suspended" from a synod of the Presbyterian Church in 1873 because of their granger activities.[31]

Grange leaders knew of these organizational problems connected with secrecy, but they were unwilling to sacrifice this feature of the society in order to make it more acceptable to a small minority. Instead of conceding to the reservations of a few churchmen and their followers, order spokesmen tried to persuade the concerned that secrecy did not contradict religious teachings, and they attempted to show them why it was expedient to keep the work of the society confidential. The arguments given in behalf of the Patrons of Husbandry by an anonymous Iowa granger writing in the *Rural Sun* were typical of those developed by the order's defenders: "Those illustrious men [the Church fathers] who paved the way for the great unbounded religious liberty we enjoy, oftentimes, for security of the word of God and their own lives, hid that greatest of all books in secret places, and bound themselves in secret orders to meet in caves and dens to counsel together." This unknown Iowan then expounded on the farmers' need of secrecy in order to protect themselves from their alleged enemies, the conspiratorial and powerful monopolies and railroads.[32]

Other religious bodies saw nothing evil in secrecy and therefore found no grounds for condemning the order's activities. In fact cordial relationships developed between many country churches and nearby subordinate granges. It was not uncommon for Protestant ministers to respect grangers and to cooperate with them in their

31 Garnett, Kansas, *Weekly Journal*, March 20, April 18, July 8, 1874. Clippings are in the James Hanway Scrapbook, VI, 1-2, 4, 9, Kansas Historical Society, Topeka (hereinafter cited as Hanway Scrapbook); Madison *Western Farmer*, April 26, 1873.
32 Nashville *Rural Sun*, February 29, 1873; Charleston *Rural Carolinian*, April, 1872.

drives. The bond between the two stemmed in part from the preachers' admiration for the agrarian group's vehement defense of blue laws and its active fight for prohibition of the alcohol traffic. Conspicuous injections of prayer and Scripture into grange meetings also convinced many persons that the order was not guilty of apostasy.[33]

Many local granges openly solicited the friendship of area churches, and congregations often reciprocated by extending brotherly assistance. In Kentucky members of the South Union Church permitted the Church Hill Grange to assemble in their building for four years, until grangers collected enough money to construct their own temple. The patrons expressed their gratitude by contributing to the congregation's "carpet fund." [34]

In addition to church opposition, grange deputies faced other problems when dealing with immigrants. Many aliens could not speak English, and the language barrier created an impasse. National and state granges were not prepared to translate rituals, proceedings, constitutions, and circulars into every tongue represented in rural America. A few attempts, however, were made to reach major non-English-speaking elements. In 1874 the Wisconsin State Grange approved a resolution authorizing the printing of its constitution in German and Norwegian, and two years later the Louisiana State Grange accommodated the state's Creole population by publishing its ritual in French.[35]

Probusiness publishers usually disagreed with grange objectives, and they often did all in their power to discredit the organization. Columns of their publications frequently carried verbal assaults calculated to affect the order's recruitment program and to place officers of the Grange on the defensive. In an era when inflammatory and libelous news reporting was commonplace, journalists and editors like E. L. Godkin of *The Nation* wrote vicious articles dealing with the order. Convinced that "it was perfectly clear that the Granger movement was rank communism, and its success in this country was against all reason and experience," Godkin tried to dishonor the organization by charging that it had been inspired by radical revolutionaries.[36]

[33] New York State Grange, *Proceedings*, 1891, pp. 130–31; 1895, p. 163.

[34] Church Hill Grange, Secretary's Minutes, May 21, 1874, December 27, 1878; Fort Smith (Ark.) *Herald*, February 28, 1874.

[35] Humboldt (Tenn.) *Grange Journal*, July 28, 1876; Chicago *Prairie Farmer*, June 3, 1876; Wisconsin State Grange, *Proceedings*, January, 1874, pp. 13, 24.

[36] "The Granger Collapse," *Nation*, XXII (January 27, 1876), 57–58. See also Charles F. Adams, Jr., "The Granger Movement," *The North American Review*,

Although such criticism hurt the image of the Grange in the eyes of the urban intelligentsia, it did little to influence the great masses living in rural areas. They seldom read city newspapers and magazines, and even if they did, they refused to accept the charges at face value. To them news of the order carried by metropolitan papers was only big city propaganda written to frighten farmers into complacency.

Far more distressing were the uncomplimentary comments penned by disgruntled grangers for publication in farmers' journals and the critical editorials carried by these magazines. Patron Charles Whitaker of Hamilton County, Iowa, wrote an article for the Chicago *Tribune* which was reprinted by several agricultural periodicals. The Iowan charged that the order had been created solely to enrich men like Kelley; he supported his accusations with the order's official financial statements for 1872. Editor George T. Anthony of the *Kansas Farmer* was of the same opinion. In one of his editorials, he called the order a "money making scheme." [37]

Damaging indictments such as Whitaker's and Anthony's usually brought prompt rebuttals from order spokesmen. In 1875 when the *Western Rural* printed a letter from a dissatisfied patron, Missouri State Grange Master Thomas R. Allen replied vigorously. Implying that the Grange was above criticism, Allen labeled the anonymous critic a "harpy" and he discounted the validity of the granger's complaints, calling them mere "twaddle." [38]

Despite charges and countercharges by men like Whitaker and Allen and in spite of debates over the theological implications of secrecy, the order grew rapidly in the Middle West and South and more slowly in the East and West during the first five years of the 1870s. Seeing the Grange as a panacea for their economic and social maladies, farmers and their wives joined the order by the thousands.[39]

CXX (April, 1875), 394–424; Marie Howland, "The Patrons of Husbandry," *Lippincott's Magazine*, XII (September, 1873), 338–42; Chicago *Tribune*, 1873–75, *passim*; Topeka *Commonwealth*, September, 1883.

37 Madison *Western Farmer*, February 21, 1874; Leavenworth *Kansas Farmer*, August 15, 1872. For another editorial attack, see Madison *Western Farmer*, March 26, 1870.

38 Chicago *Western Rural*, June 19, 1875. Another series of charges and defenses was carried in the Athens (Ga.) *Southern Cultivator*, July, September, October, 1873.

39 National Grange, *Proceedings*, February, 1874, p. 21; February, 1875, p. 22; November, 1875, p. 190; 1876, pp. 178–79; Topeka *Kansas Farmer*, March 11, 1874; Robert L. Tontz, "Memberships of General Farmers' Organizations, United States, 1874–1960," *Agricultural History*, XXXVIII (July, 1964), 154.

The following table sheds light on the geographical nature of the First Granger Movement. The popular notion that the center of the movement was located somewhere in the upper Mississippi River Valley because the legislatures of Iowa, Illinois, Minnesota, and Wisconsin passed railroad regulator laws in the 1870s cannot be substantiated by membership records. Instead, they refute that contention. Iowa was the only state of the group ranked among the top ten in membership.

Table 6

GRANGE MEMBERSHIP, 1875

State	No. of Granges Organized	Number Still Active	Member-ship
Alabama	678	531	17,440
Arkansas	634	631	20,471
California	263	263	14,228
Connecticut	16	16	480
Delaware	23	23	503
Florida	143	83	3,804
Georgia	708	545	17,826
Illinois	1,592	788	29,063
Indiana	2,036	1,485	60,298
Iowa	2,004	1,162	51,332
Kansas	1,373	406	40,261
Kentucky	1,618	1,545	52,463
Louisiana	316	315	10,078
Maine	189	189	8,247
Maryland	158	152	5,635
Massachusetts	100	98	3,825
Michigan	609	605	33,196
Minnesota	546	455	16,617
Mississippi	669	645	30,797
Missouri	2,034	1,901	80,059
Nebraska	620	288	8,177
Nevada	15	15	378
New Hampshire	69	69	2,528
New Jersey	96	94	4,495
New York	354	273	11,723
North Carolina	540	342	10,166
Ohio	1,216	878	53,327

Table 6— (Cont.)
GRANGE MEMBERSHIP, 1875

State	No. of Granges Organized	Number Still Active	Member- ship
Oregon	186	186	8,233
Pennsylvania	615	534	22,471
Rhode Island	0	0	0
South Carolina	357	341	10,922
Tennessee	1,097	1,091	37,581
Texas	1,210	1,201	37,619
Vermont	207	198	10,193
Virginia	670	663	13,885
West Virginia	312	280	5,990
Wisconsin	514	446	17,226
Arizona	0	0	0
Colorado	69	63	2,098
Dakota	56	53	1,178
District of Columbia	1	1	45
Idaho	16	16	390
Montana	26	26	946
New Mexico	0	0	0
Indian Territory	15	15	450
Washington	66	66	2,169
Canada	27	27	450
Total			761,263 [40]

Solon Buck perpetuated this misconception by devoting so much of his book to granger activities in the upper Mississippi River Valley and by equating the so-called granger laws with granger strength. Moreover, Buck used the word *granger* to describe any farmer dissatisfied with the status quo. In other words he lumped all dissenting farmers into one category, whether they affiliated with the Grange or not. Taken in this sense, the First Grange Movement might easily have had its center somewhere in the upper Mississippi River Valley. In Illinois alone there were approximately twelve hundred granges and two thousand independent farmers' clubs in operation by early 1874.[41]

If, on the other hand, the word granger is used to designate only

40 National Grange, *Proceedings*, November, 1875, p. 190.
41 Buck, *The Granger Movement*; Chicago *Industrial Age*, April 18, 1874.

members of the Patrons of Husbandry, Buck's emphasis was entirely without statistical justification. Membership records show that the real source of grange numerical power rested in the Ohio River Valley and in Missouri, not in the upper Mississippi River Valley. Roughly one-third of the order's members resided in Ohio, Indiana, Kentucky, and Missouri. Action taken by the National Grange in late 1875 reflected the growth of the order in the Ohio River Valley. After screening many cities, the executive committee urged the National Grange to move its headquarters from Washington, D.C. to Louisville because the Kentucky metropolis was in the middle of the granger belt.[42]

The society was well received in other areas as well. Kansas, Texas, Tennessee, Michigan, and Mississippi grangers accounted for almost one-fourth of the national membership. In general grange popularity in 1875 stretched from Pennsylvania to the Great Plains and from Virginia to Texas. The order also prospered in California, but it did not fare very well in New England and New York.

Nor were membership statistics the only measure of the impact of the First Granger Movement. There were other signs pointing out the order's success. When the organization was at its zenith in the mid-1870s, merchants began wooing the patrons' business. In Nemaha County, Nebraska, entrepreneurs revamped their advertising to cater to the weaknesses of unsuspecting grangers. The *Nebraska Advertiser* carried paid notices for "Granger Cigars" and the "Granger Saloon." The proprietor of the latter establishment announced that he sold the "best of liquors provided at Grange prices" and inserted a popular order slogan, "Down with monopoly," to convince farmers that he was solidly behind their cause.[43]

In contrast to most antebellum agricultural clubs, the Grange was not class oriented. Order rules permitted all farmers and their wives to be inducted into the society, and judging from available data the organization represented a cross-sampling of American rural life. In most communities, old and young members of both sexes and from all economic levels massed together to work for common goals. It was not uncommon for prominent men like Governors E. F. Jones of New York and W. R. Taylor of Wisconsin and State Commissioner of Agriculture Leonidas L. Polk of North Carolina to cooperate fully with dirt farmers in grange halls across the country; nor was it out of the

[42] National Grange, *Proceedings*, November, 1875, pp. 19–20.
[43] Brownville *Nebraska Advertiser*, April 9, 1874.

ordinary for teenagers and men and women in their sixties, seventies, and even eighties to belong to the same chapter.[44]

Of course the well-known men who belonged to the order often received special honors. They frequently received invitations to speak at grange "pick nicks" and to officiate at lodge hall dedications. Whenever grangers assembled for a specially planned meeting, they usually wanted an outside speaker to add prestige to the event. Grange celebrities like L. L. Polk were besieged with requests to address these gatherings and to revive interest in the order.[45]

Although grange membership was theoretically open to all farmers, there were still large blocs that were rejected openly or otherwise. For instance, the constitution of West Butler Grange No. 476 of Nebraska rigidly excluded "any member found guilty of wanton cruelty to animals, or gross immoral conduct, or not living in obedience to the laws of the state." [46]

Negroes were the largest excluded group. The National Grange generally ignored the problem of discrimination and refused to censure subordinate and state chapters which barred minorities on racial grounds. As a result, enterprising southern whites were able to deny membership to blacks by developing for them an allied but separate group known as the Council of Laborers. Thus, southern patrons always had an excuse for excluding black farmers, although the Council did not succeed. In Mississippi, the Grange was often a front for the Ku Klux Klan. The farmers' organization with its features of secrecy and brotherhood and its legitimate objectives was subverted to serve as a cloak for the devious work of recalcitrant opponents of Reconstruction. There were, however, some exceptions of southern granges not warped by prejudices against Afro-Americans. A few locals in Louisiana were known to accept applicants "without regard to color." The two races in these chapters, according to one witness, worked har-

[44] Selected letters to Leonidas L. Polk, 1875–78, in Polk Papers, Southern Historical Collection, University of North Carolina, Chapel Hill; Charleston *West Virginia Farm Reporter*, December, 1897; Madison *Western Farmer*, April 19, 1873; Nashville *Rural Sun*, May 15, 1873; Social Grange (Ill.) No. 1308 Roll Book, 1874–1904, Social Grange Records, Illinois State Historical Society, Springfield; Winnebago County (Ill.) Pomona Grange, Membership Applications, 1888–99, Winnebago County Pomona Grange Records, Illinois State Historical Society, Springfield; Massachusetts State Grange, *Proceedings*, 1898, pp. 91, 93.

[45] D. A. Montgomery to Polk, July 22, 1877, J. E. Porter to Polk, July 26, 1877, March 18, 1878, E. J. Brooks to Polk, August 7, 1877, O. L. Ellis to Polk, August 28, 1877, C. D. Rountree to Polk, June 3, 1878, all in Polk Papers.

[46] West Butler Grange (Nebr.) No. 476, Constitution and By-Laws, West Butler Grange Records, Nebraska Historical Society, Lincoln.

moniously "on a basis of mutual interest and common defense of the farming classes against the political jobbers and monopolists." [47]

By its very nature the First Granger Movement had the seeds of its own destruction. Rather than promoting brotherhood, secrecy often caused dissension and mistrust. When the order failed to fulfill all of its promises, many members became suspicious and began accusing each other of conspiring against the organization and of cheating. Even grange stalwarts like Oliver H. Kelley were not immune from investigations for fraud. After he had tendered his resignation November 21, 1878, the executive committee thoroughly studied his records to ascertain whether or not he had stolen from the society. To protect himself Kelley had to account for every item belonging to the national secretary's office.[48]

The order was fully prepared to deal with ordinary members' misconduct. Grange bylaws defined behavior that constituted sufficient grounds for suspending, expelling, or fining a member for breaking the rules of the organization. The bylaws of Cedar Mountain Grange No. 353 of Virginia divided improper conduct into four categories. Under the heading of "irregular practices," patrons faced expulsion if they were found guilty of "habitual drunkenness or other immoral conduct." Using profanity, being cruel to animals, and disobeying the master of the grange came under the class of "disorderly behavior," while divulging society secrets and bringing malicious lawsuits against fellow members were two cardinal offenses outlined by the third grouping. The fourth category embraced absenteeism and provided penalties for missing sessions. It was the most commonly used. According to the bylaws, if a member was not present for three consecutive assemblies of the chapter without an acceptable excuse, he received a fine of twenty-five cents.[49]

The judicial system of the Grange was quite democratic. Members accused of breaking serious regulations of the order were tried by a grange tribunal. At the local level, subordinate lodges had jurisdiction over members, but appeals for new trials could be made to the state or national grange. Litigation in order courts generally followed the same legal procedures as in American civil courts. They heard testimony, listened to evidence, reached verdicts after both sides had

[47] Theodore Saloutos, "The Grange in the South," *Journal of Southern History*, XIX (November, 1953), 476–78.
[48] National Grange, *Proceedings*, 1878, p. 37; 1879, pp. 106–107, 145.
[49] Cedar Mountain (Va.) Grange No. 353, *By-Laws* (Culpeper, 1875), 13–14.

presented their cases, and issued sentences. The only major difference was that the membership of the grange served as both judge and jury.[50]

Most of the trials conducted by granges involved corrupt officers and delinquent chapters. Minor offenses such as not paying dues or not attending meetings usually did not warrant a trial. One of the most celebrated grange trials involved a questionable chapter chartered in Boston by Deputy J. C. Abbott. The controversy began when national leaders discovered that the Massachusetts local was composed of grain dealers, commission men, a reporter for a business journal, and an editor of a political newspaper. After much controversy, the national body ruled that the subordinate chapter had to surrender its dispensation. Bay State patrons resented the expulsion order, and so friction developed between them and the National Grange when they refused to enforce the grange high court's edict. Before the conflict was finally settled, the National Grange had to oust the illegal Boston chapter and censure the state body for its insubordination.[51]

Grangers also lodged many complaints against inept and voracious officers. Among the many cases which arose in Nebraska was one resulting from charges by State Secretary L. C. Root against Master J. F. Black of Red Willow Grange No. 28. Root claimed that Black had made "malicious and unjust charges against" him and other members of the state grange and had "caused to be bublished [sic] in the Indianola Courrier [sic] . . . a certain resolution purporting to have passed the State Grange." The state executive committee investigated the case and found Black guilty of slander and of issuing false news releases concerning the order's activities. For his misbehavior, Black received a one-year suspension from the organization.[52]

Executive committees occasionally investigated reports of wrongdoing and found no evidence of misdeeds. In South Carolina Master Paul Livingston of Witts Mills Grange freed himself of accusations of permitting nongrangers to observe secret proceedings by proving that the men in question were actually active members of the order. Another unfortunate incident occurred in Georgia. Rumors circulated in grange circles in 1875 that State Agent T. G. Garrett was neglecting his duties. The case was somewhat unique, however, because Garrett

---

[50] National Grange, *Proceedings*, 1875–99.

[51] Charleston *Rural Carolinian*, April, 1874; Chicago *Industrial Age*, November 1, 1873.

[52] General Records of the Secretary of the Nebraska State Grange, in Nebraska State Grange Records, Nebraska Historical Society, Lincoln.

quieted his critics himself without trial by offering "to give every man his due . . . if any brother has been wronged." [53]

Grange trials undoubtedly brought disunity to the order and had a bearing on the organization's decline in the 1870s. Accusations and denials divided "brothers" and "sisters" into rival factions. Some members defended the accused, while others sided with the complainant. If tempers became heated, as they did on occasion, relations would rupture, and the society would split in half. Of course it is impossible to measure with any degree of accuracy the influence that trials and hoaxes played in terminating the First Granger Movement, but it is safe to state that judicial proceedings, whether based upon fact or hearsay, did their part to undermine the movement.

Had there never been grange purges, the First Granger Movement would have lost its power because other factors were at work to weaken the organization. Many patrons were fickle. They quit paying their dues and stopped attending meetings after the initial excitement of being associated with a secret fraternal organization had worn off. For these lukewarm supporters, grange ritual quickly became boring, and their interests wandered in other directions. Members' correspondence and grange records attest to the fact that many farmers and their wives left the order without much explanation soon after they had joined it.[54]

Others quit the order because they believed that state and national leaders squandered their dues. Money flowing into state and National Grange coffers was used almost exclusively to pay travel and per diem expenses of delegates attending order conventions and to provide salaries for a select group of officers. Actions taken in this regard by members of Chippewa Grange No. 120 of Wisconsin, December 18, 1875, reflected dissatisfaction with wastefulness. These Wisconsinites showed their displeasure with the appropriation of their dues for ex-

53 B. F. Barton to D. Wyatt Aiken, December 31, 1875, in Clemson University Patrons of Husbandry Papers, Clemson University Library, Clemson, S.C.; Georgia State Grange, *Proceedings*, 1875, pp. 8–9, 13–14.

54 Elizabeth Benjamin to John Benjamin, June 11, 13, 1875, in John Benjamin Family Papers, Minnesota Historical Society, St. Paul; James C. Hope to the master of the Pomaria Grange, January 19, 1878, in Clemson University Patrons of Husbandry Papers; Louisville *Farmers Home Journal*, February 14, March 21, 1878; Olive Grange (Ind.) No. 189, Secretary's Minutes, December 17, 1885, January 9, 1887, in Olive Grange Records, Indiana Historical Society, Indianapolis; Church Hill Grange, Secretary's Minutes, December 17, 1885; Atlanta *Georgia Grange*, June 30, 1877; Raleigh Grange (N.C.) No. 17, Secretary's Minutes, April 4, 1874, in Duke University Patrons of Husbandry Papers, Duke University Library, Durham.

cessive expense accounts and salaries by voting unanimously to make
a formal protest to the state grange. The state body evidently saw no
merit in the Chippewa grangers' request; its officers refused to trans-
mit its resolutions to the national conclave. Their refusal to take a
more cost-conscious stand no doubt contributed to the disbandment
of the Wisconsin local in 1876.[55]

Human nature predicated that every time state or National Granges rejected some pet program, undercurrents of dissent would dis-
rupt the harmony needed for the successful operation of the order.
Although the editors of grange proceedings generally avoided controversy and refused to print pungent descriptions of floor debates, battles over dialectic issues must have ensued and caused some degree
of impenitence. How much polemical discussions contributed to the
rapid withdrawal of members from the organization cannot be accurately determined, but no doubt stubbornness and inability to accept defeat were vital factors in the termination of the First Granger
Movement.[56]

Some impatient patrons resigned from the Grange in order to join
the more militant Farmers' Alliance. The fraternal order moved too
deliberately and too indecisively to satisfy these agrarian activists,
while the Alliance promised to tackle problems more directly. Moreover, the Grange tried to pursue a neutral political course, but the
other body made no attempt to shun partisanship. In many areas the
older organization waged a desperate battle against insurmountable
odds to persuade its members to remain loyal to the order.[57]

Ironically, another group of members quit the Grange because they
thought the order had already fulfilled its mission. State and national
lawmakers were gradually becoming more and more reform-minded,
and the nation's farmers had demonstrated that they were able to discard their individuality enough to form an effective lobby. Many
grangers looking at the prevailing situation surmised that victory
had been won. To them the Grange had outlived its usefulness, and

[55] Conclusions based upon treasurers' financial reports found in state and national grange proceedings listed in the bibliography; Wisconsin State Grange, *Proceedings*, 1884, p. 29; 1876, p. 7; 1877.
[56] For evidence of disagreement, see Wisconsin State Grange, *Proceedings*, 1880,
pp. 69, 73; 1884, pp. 28–29; Illinois State Grange, *Proceedings*, 1899, pp. 43–46.
[57] Augusta *Southern Cultivator and Dixie Farmer*, December, 1888; Ralph A.
Smith, "The Grange Movement in Texas, 1873–1900," *Southwestern Historical
Quarterly*, XLII (April, 1939), 303.

they, therefore, saw no reason why they should not drop out of the organization.[58]

Some patrons protested order activities and actions by resigning from the organization. Grange founder William Saunders severed his ties with the fraternal society November 24, 1875, because it was transferring its national base of operations from Washington to Kentucky. Although he contended that his busy schedule did not permit him time "to reach and stay at Louisville for even a day," his wounded pride probably had more to do with his resignation than the inconvenience caused by the shift. The pioneer granger had valiantly pleaded in vain with fellow National Grange members not to move the offices. After being conspicuously absent from the organization for five years, Saunders' vexations dissipated, and in 1880 the backslider returned to the fold.[59]

The Saunders case was not unlike one involving Kelley. After alluding to the fact that for some time he was planning to relinquish his duties as a national officer, Oliver H. Kelley finally tendered his resignation as secretary of the National Grange November 21, 1878. Although the founder stated that his decision had been based upon a desire to devote full time to land speculation in Florida, his disagreements with the executive committee most likely prompted his decision. Having on occasion referred to the members of that body as "an expensive ornament," Kelley subtly manifested his disdain for them in his correspondence. Evidently the feeling was mutual because upon his resignation, the committeemen launched a thorough investigation of Kelley's operations as secretary and checked carefully to see if he had cheated the organization in any way. If prominent men like Kelley and Saunders left the order out of protest, certainly countless others must have done the same thing.[60]

The Grange naturally attracted some opportunists who expected to gain rich rewards from membership. Many aspiring farmer-politicians like Ignatius Donnelly of Minnesota hoped to use their affiliation with the order as launch pads to public office. After their motives for join-

[58] Atlanta *Georgia Grange*, June 30, 1877; Wisconsin State Grange, *Proceedings*, 1879, p. 14.

[59] National Grange, *Proceedings*, November, 1875, pp. 88–89; 1880, pp. 5–13.

[60] *Ibid.*, 1878, p. 37; 1879, pp. 106–107, 145; Oliver H. Kelley to S. H. Ellis, November 4, 1875, July 15, 1876, August 7, September 5, 1877, in Louis J. Taber Papers, Grange Historical Material, Collection of Regional History, Cornell University Library, Ithaca, N.Y.

ing the society had gained them the goals they sought and after the First Granger Movement had ceased to be a powerful force in rural America, they left the order. Another group of selfish men became members to take advantage of low prices offered by grange cooperative enterprises. Patrons of this stripe gave little and got much from the organization. They vanished quickly after mutual business ventures began failing.[61]

In fact, as Solon Buck ably pointed out, the demise of cooperatives had an important bearing on the order's decline. But he failed to recognize the domino effect that this factor had on the First Granger Movement. Did the failure of business enterprises precede the order's decline in membership or did the other factors causing deterioration come first? In the final analysis there is some truth in both hypotheses. Grange cooperatives needed customers to keep operating, but many times the businesses' inexperienced managers drove away their clientele and exhausted treasuries by making costly mistakes. When this happened, members who had invested their money in these fledgling enterprises became impatient and quit the order. But occasionally cooperatives were run efficiently with no problems until members abandoned the Grange. In these cases, the order's decline caused mutual businesses to go bankrupt.[62]

Buck cited one factor which had little effect on the collapse of the First Granger Movement. He claimed without adequate verification that "the connection of the Grange with a number of political movements of the time contributed in large measure to its decline." The organization's relationship with political parties was almost always one of neutrality, not partiality. Most granges adhered to the order's rule against taking sides in elections. Of course there were some exceptions, but state leaders usually chided noncomformists for actively backing slates of candidates. Since incidents of this type were so rare, it would be a gross exaggeration to attach as much significance as Buck

---

[61] For discussions of Donnelly's career as a granger, see Martin Ridge, "Ignatius Donnelly and the Granger Movement in Minnesota," *Mississippi Valley Historical Review*, XLII (March, 1956), 693–709; and Martin Ridge, *Ignatius Donnelly: The Portrait of a Politician* (Chicago, 1962). A detailed description of Donnelly's sordid political ambitions will be given in a later chapter. Atlanta *Georgia Grange*, June 30, 1877; Michigan State Grange, *Proceedings*, January, 1875, p. 6; Nebraska State Grange, *Proceedings*, 1874, p. 7.

[62] Buck, *The Granger Movement*, 73. A detailed treatment of grange cooperatives will be found in Chap. 7.

did to the correlation of politics and the mass withdrawal of members from the Grange.[63]

Rapid growth during the First Granger Movement, which exceeded the most sanguine hopes of the order's founders, caused unforeseen problems. Kelley and his associates were not prepared to deal with "an unwieldy and undisciplined mass of members." As a result, intercourse between granges was limited, and individual chapters did not benefit from exchanges of ideas. Had there been adequate machinery and manpower to provide informative newsletters about the activities of local lodges, much of the dullness of grange meetings could have been alleviated. As Master Thomas Taylor of the South Carolina State Grange suggested in his annual address in 1875, it was expecting too much from local leaders relying almost solely upon their own imaginative skills to ask them to provide stimulating programs every time patrons assembled. Since the challenge was too great for many grangers to overcome, they gave up and disbanded their chapters.[64]

Although there were no provisions in order constitutions and by-laws for disbanding granges, most locals which passed out of existence probably followed the same general procedure as did North Star Grange No. 1 of Minnesota. On September 21, 1878, members of this pioneer subordinate lodge voted to "*Disband* and dispose of the personal effects equally amongst its members in good standing." Property belonging to the chapter was then auctioned off to the highest bidders, and the proceeds from the sale were divided equally among members. Occasionally, weak locals were absorbed by stronger chapters.[65]

The Grange survived tremendous membership losses and began to revive in the interim between the two granger movements because a small group of dedicated Patrons of Husbandry refused to abandon the organization. Their determination to rebuild the order reflected itself in many ways. At the national level, delegates continued to convene each year to discuss problems related to the organization's structure and to revise the order's ritual. Their critical examinations of the Grange uncovered one major flaw in the society's program. They found that not enough young adults had joined the order to compensate for losses incurred by older members' deaths. They attributed this

63 *Ibid.*, 72. The exceptions will be discussed in Chap. 8.

64 *Ibid.*, 71; South Carolina State Grange, *Proceedings*, December, 1873.

65 North Star Grange (Minn.) No. 1, Secretary's Minutes, September 21, 1878, in North Star Grange Records, Minnesota Historical Society, St. Paul; Church Hill Grange, Secretary's Minutes, February 26, 1875, March 10, 1876.

phenomenon to the fact that no provision had been made to interest farmers' children in the organization. At the 1888 session of the National Grange, this problem was not only discussed, but steps were taken to remedy the situation. The delegates voted to study a Texas State Grange proposal for juvenile granges. After carefully evaluating the advantages and disadvantages of the Texans' scheme, the National Grange adopted the plan for use in all local chapters.[66]

Local and state leaders also tried to save the order from extinction. Their determined efforts, however, were not always enough to revive interest in the organization. Such was the case in Kentucky. Even after the National Grange had denied the state representation at the 1878 session for failing to pay its dues for four consecutive quarters, Master W. J. Stone was still confident that he could make the organization financially solvent again in the Blue Grass State. Although it took over a year of frustration to convince him that he had misjudged the desires of Kentucky farmers, Stone finally accepted the hopelessness of the situation and reluctantly disbanded the state grange in 1880.[67]

Illinois grangers used various methods to stave off apathy. In 1875 the officers of Champaign Grange No. 621 developed a plan for reviving interest in their area and for collecting unpaid dues. They divided the county into three sections and assigned missions to visit delinquent members in each division. Although this scheme did not have a permanent bearing on the order in this central Illinois county, it nevertheless showed a willingness on the part of local grange leaders to experiment with different proposals to prolong the life of the order in their communities. Officials of the Illinois State Grange attempted to inject vitality into the order in the early 1890s by offering a banner to the county most active in securing new recruits.[68]

The persistence of grangers in keeping the order alive finally paid dividends in the late 1880s and 1890s. Signs pointing to a revival and to a second granger movement were gradually beginning to appear in the East even while the order was declining in other parts of the coun-

[66] J. D. Whitham to John P. Clarke, October 22, 1879, in John P. Clarke Papers, West Virginia University Library, Morgantown; Augusta Southern Cultivator and Dixie Farmer, December, 1888; National Grange, Proceedings, 1888, pp. 99–100; California State Grange, Proceedings, 1891, pp. 13–14; New York State Grange, Proceedings, 1877, pp. 78–79.

[67] Louisville Farmers Home Journal, January–December, 1879.

[68] Champaign Grange (Ill.) No. 621, Secretary's Minutes, November 15, 1875, in Champaign Grange Records, Illinois Historical Survey Collections, University of Illinois Library, Urbana; Illinois State Grange, Proceedings, 1890, pp. 29–30; 1891, p. 79; 1894, p. 20; 1895, p. 18.

try. A certain sense of urgency brought on by increasing competition and falling prices was developing among eastern farmers. Their land tended to be rocky and their soil fertility depleted, and their holdings were often too small and too hilly to permit them to use modern large-scale agricultural methods. Disadvantageous conditions also prevented many eastern farmers from trimming expenses enough to meet the stiff challenge of western foodstuffs. As a result a climate of fear gripped the whole area. Those unable to compete abandoned their farms and moved to town or to the West, while many of the better situated agriculturists in the East looked to the Grange for relief.[69]

Grange membership records leave no doubt that unfavorable eastern agricultural conditions had an impact on the organization. The order grew very rapidly throughout New England, New York, Pennsylvania, and Ohio during the last two decades of the nineteenth century.

Table 7

FAMILY MEMBERSHIPS IN THE GRANGE

State	1875	1880	1890	1900
Ohio	28,067	12,317	4,418	10,354
Maine	4,340	3,137	7,984	12,805
New Hampshire	1,330	767	7,988	11,762
Vermont	5,365	1,419	903	1,063
Massachusetts	2,013	628	2,061	6,913
Connecticut	253	0	4,493	3,465
Rhode Island	0	0	509	587
New York	6,407	3,110	8,438	22,715
Pennsylvania	11,827	5,964	6,661	12,335
Regional Total	59,662	27,342	43,455	81,999
U. S. Total	451,605	65,484	71,295	98,675[70]

The preceding chart points out clearly the regional character of the Second Granger Movement. At the crest of the first grange membership drive in 1875, only 13 percent of the Patrons of Husbandry lived in New England, New York, Pennsylvania, and Ohio, but the percentage increased to 41 in 1880. Ten years later, three-fifths of all

[69] Maine State Grange, *Proceedings*, 1897, pp. 8–9; Massachusetts State Grange, *Proceedings*, 1899, pp. 74–75.

[70] Tontz, "Memberships of General Farmers' Organizations," 154.

grangers resided in this region, and at the turn of the century 83 percent of the membership was found in these nine states.

These regional trends of grange strength account for only one major difference between the first and second granger movements. Annual grange membership statistics for individual states highlight one other factor about growth patterns which distinguishes the two periods. The influx of farmers and their families into the Grange in the 1870s was very rapid, but growth in the East during the 1880s and 1890s was very gradual. New Hampshire, New York, and Massachusetts, like the other five states in the region, illustrated the trend from 1881 to 1899.

Eastern farmers had the same basic motives in mind when they joined the order in the 1880s and 1890s as had those who affiliated with the society in the 1870s. Both groups expected economic, social, and educational improvements to result from their membership in the Grange. As a result of this common feeling, the organization's pro-

Table 8

GRANGER ACTIVITY IN NEW HAMPSHIRE

Year	No. of Granges	Members
1883	64	3,443
1884	73	3,973
1885	77	4,422
1886	86	4,983
1887	92	5,865
1888	103	6,701
1889	107	7,560
1890	122	8,838
1891	130	9,870
1892	... No Report	.....
1893	173	13,242
1894	189	14,832
1895	...	......
1896	... No Report	......
1897	...	......
1898	238	20,720
1899	255	22,330[71]

71 New Hampshire State Grange, *Proceedings*, 1891, p. 38; 1893, p. 11; 1894, pp. 42–43; 1898, pp. 29–30; 1899, p. 36.

Table 9

GRANGER ACTIVITY IN MASSACHUSETTS

Year	No. of Granges	Members
1881	33	1,141
1882	37	1,529
1883	40	2,088
1884	45	2,696
1885	49	3,335
1886	55	3,938
1887	65	4,723
1888	97	6,267
1889	..	No Report .....
1890	112	8,964
1891	111	9,374
1892	119	10,230
1893	120	10,719
1894	125	11,400
1895	129	11,611
1896	134	12,238
1897	141	12,810
1898	146	13,177
1899	146	13,137[72]

Table 10

GRANGER ACTIVITY IN NEW YORK

Year	Net Membership Gains
1889	3,458
1890	8,036
1891	4,144
1892	1,298
1893	1,509
Total Net Gains: 1889–93—18,445	
Total Membership: 1893—33,584[73]	

[72] Data gathered from statistical reports given in Massachusetts State Grange, *Proceedings*, 1882–1900.

[73] Information tabulated from records in New York State Grange, *Proceedings*, 1890–94.

grams did not change very much in the nineteenth century, and patrons' demands during the two granger movements did not differ noticeably. No special point will be made therefore in this study to differentiate between the two phases of the movement.

CHAPTER III

The Grange
and Lower Education

An assessment of the role exerted by the Grange in shaping late nineteenth-century agrarian life and policies has traditionally meant one of two things to most students of the period. The organization has either been credited with uniting rural antimonopoly sentiments into a protest against railroads and middlemen or it has been linked with allied farmers' actions against political ineptitude and irresponsibility. This characterization has the order associated with the formation of various independent and antimonopoly political leagues which sprang up spontaneously in the 1870s and 1880s and with the development of the Greenback and Populist parties. Grange meetings have thus been pictured as assemblies of disgruntled and frustrated country folk brooding over strategies for unseating their oppressors. According to this popular interpretation of the order's place in history, crusades against corporate and entrepreneurial wealth and battles with political corruption and legislative stagnation were the only sides of grangerism worthy of mention; all other aspects of the movement were secondary and largely unimportant.

Solon Buck perpetuated this narrow view by devoting approximately 95 percent of *The Granger Movement* to the political and economic activities of the organization.[1] His monograph lacks balance because he slighted other parts of the order's program. In actuality the educational and social features of the society were the order's most important activities. Not only did more grangers participate in them, but these features produced more lasting results.

When Kelley investigated southern agricultural conditions in 1866, the inadequate educational opportunities offered rural children and their parents' lack of interest in this vital matter concerned him. He resolved to channel farmers' negativism into positivism and to find formulas for enhancing their lot. To him the infinite powers of education were the keys which removed the chains holding farmers and

[1] Buck, *The Granger Movement.*

their children in a helpless servile position. Kelley's convictions were
so strong that he had no trouble persuading his Washington associates to include provisions for educational improvement in the order's initial plans and to emphasize its paramount importance in the
Grange's overall program.

Thus, besides aiding farmers economically and politically, the
Grange also was interested in helping their children. Specifically,
grangers undertook to give their offspring better educational opportunities, a new attitude since the average farmer tended to think "that
education and farming, like oil and water, would not mix; that it was
impossible for him [or his children] to use the educational advantages
within reach of others." [2] Before the advent of the Grange, in fact,
most farmers concluded that the only worthwhile lessons that the
country school offered its students were the "three R's," while the remainder were not interested in even that limited goal. But later, many
farmers who belonged to the organization recognized that an education could be valuable to rural youths, provided that it was tailored to
their specific needs.

Everywhere, state and national grange committees and order
spokesmen regularly subjected ordinary members to exhortations
about the virtues of learning. In 1882 Texas grangers heard their
state master, A. J. Rose, say that "nothing increases debt, vice, super-
stition, and crime so much as ignorance" and remind them not to forget "that education is the leading feature of the order." Eight years
later the National Grange committee on education asked grangers to
stay abreast of their states' school laws in order "to keep them up with
the progressive times in which we live." [3] Literally hundreds of similar appeals were made in the late nineteenth century, and they caused
farmers to consider the possible benefits accruing from "book learning" instead of the ageless assertions regarding the impractical nature of education.

A purely academic quest of knowledge, however, was not necessarily the objective of grangers. They usually had different motives
for supporting education, and they almost always wanted it to serve
them in a special way. If end results did not match their preconceived
objectives, they generally complained and sought alterations.

In its campaign for educational improvement, the Maine State

2 Arthur E. Paine, *The Granger Movement in Illinois* (Urbana, 1904), 46.
3 Texas State Grange, *Proceedings*, 1882, p. 6; National Grange, *Proceedings*,
1890, p. 151.

Grange even blamed the state's agricultural worker shortage on teachers. After criticizing "the general run of farm laborers," members of the committee on agriculture attributed the problem to educators' and parents' practice of "distilling into the minds of their children and pupils a distaste for farm life." Committeemen contended in 1898 that as long as degrading remarks about the agricultural profession circulated in classrooms and at family dinner tables, farmers would have to rely upon "dullard" helpers "whose only ambition is to obey orders and see the sun set" because not "enough of our brightest and oldest boys and girls" would remain on family homesteads to provide a talented labor supply. In this case, as in many others involving their educational demands, what grangers really wanted schools to do was inculcate pupils with a mythical grandeur of agrarian life.[4]

In the early stages of the First Granger Movement, the order did not always seek improvements in public school systems; instead, some of its leaders believed it would be easier to shape private academies to the needs of localities. This feeling was especially prevalent in the South where most public education systems originated during the hated Reconstruction Era.

The private school idea was pushed most vigorously in Alabama. In 1875 the state grange committee on education urged subordinate chapters to charter and operate their own independent schools. Consequently, four private institutions appeared in the state, including a boarding academy maintained by Cornelson Grange of De Armanville, Calhoun County, and Mountain Grange High School operated by Trinity Grange of Morgan County. Among other subjects, the latter school curriculum included astronomy, chemistry, Latin, and Greek. Two other institutions were located at Mt. Willing and at Pleasant Valley Church.[5]

The grange-inspired private school movement budded in other states, too. In North Carolina it received a boost from the legislature when that body agreed to subsidize construction of all district schoolhouses. Patrons from several communities took advantage of the law

4 Maine State Grange, *Proceedings*, 1898, p. 40.

5 John H. Franklin, *Reconstruction After the Civil War*, Vol. III in *The Chicago History of American Civilization,* Daniel J. Boorstin, ed. (Chicago, 1961), 109–11; Brownville (Nebr.) *Advertiser*, April 2, 1874; Robert Partin, "Black's Bend Grange, 1873–77; A Case Study of a Subordinate Grange of the Deep South," *Agricultural History*, XXXI (July, 1957), 55; William W. Rogers, "The Alabama State Grange," *The Alabama Review*, VIII (April, 1955), 110; William W. Rogers, "Agrarianism in Alabama, 1865–1896" (Ph.D. dissertation, University of North Carolina, 1958), 132; Alabama State Grange, *Proceedings*, 1875, p. 35.

by creating schools under lodges, "thus familiarizing the children of
the school-district with the idea that the Grange fosters education,
and that the school-house is the basis of the Grange hall." Some of
these institutions flourished as long as the order existed in the state.
One of Leonidas L. Polk's last official acts as a grange officer came June
11, 1880, when he addressed the graduates of Grange High School at
Woodland, North Carolina. There were also grange schools in South
Carolina, Mississippi, Louisiana, Michigan, and California. In the latter state, members of Franklin Grange subscribed enough money to
finance the construction of a high school at Georgetown.[6]

Moreover, South and North Carolina state granges each had unique
working arrangements with Carolina Military Institute of Charlotte,
North Carolina. Colonel John P. Thomas, the superintendent of the
academy, allotted four annual scholarships to both state executive
committees for distribution to worthy candidates. Recipients received
free tuition and a 25 percent reduction in board and room expenses.
Unfortunately, only a handful of applicants vied for the honor.[7]

As grange academies began to falter, the order launched a massive
campaign to adapt public school lessons to farm life. Oliver H. Kelley
initiated the parent organization's drive for inclusion of agricultural
subjects in elementary and secondary school curriculums at the 1877
session of the National Grange. His address advanced some revolutionary educational theories and placed him and the Grange in the
forefront as leading exponents of life-adjustment educational philosophy. Kelley's prodding of grangers "to relieve distress and avoid future troubles by encouraging the establishment of industrial schools
and making agriculture one of the principal studies" marked a major
turning point in the order's efforts in behalf of public education.
Never before had the National Grange been exposed to arguments
pointing out the need for vocational training in lower academic institutions; never had anyone actively challenged the wisdom of sub-
jecting all children to heavy dosages of the classics and the "3 R's";
and never had a high ranking figure in the organization made refer-

---

[6] National Grange, *Proceedings*, 1877, p. 129; Commencement Exercises of the
Grange High School in Woodland, N.C., B. I. Beale to Polk, June 10, 1878; both in
Polk Papers; Louis B. Schmidt, "The History of the Granger Movement," *The
Prairie Farmer*, XCIII (February 17, 1921); James S. Ferguson, "Agrarianism in
Mississippi, 1871–1900; A Study in Nonconformity" (Ph.D. dissertation, Univer-
sity of North Carolina, 1952), 177–78; Macon *Missouri Granger*, December 8, 1874.

[7] South Carolina State Grange, *Proceedings*, 1877, pp. 5–6; G. W. Lawrence to
B. C. Manly, April 24, 1874, Duke University Patrons of Husbandry Papers.

ence at a national assembly to the weaknesses of the practice of teaching all courses by rote memory. Kelley went on to say that "teaching in school should be how the scholar may earn his own living, so that when he leaves school he may enter at once upon a practical existence." [8]

The remarkable ideas expressed in Kelley's speech reflect the handiwork of a master critic and an adroit strategist who knew how to cope with the problem of correcting educational deficiencies. Kelley's keen observatory powers had enabled him to recognize the faults of the conventional approach to learning and his capacity for altering the status quo pushed the order deeper into the field of education. In fact, his dissection of the nation's traditional methods and his suggestions became the basis of the grange nineteenth-century drive for improved rural schools.

Kelley recommended two changes which would seemingly upgrade schools and make them more democratic. Since only a select group of rural students had "the means or time to secure a collegiate education" and since girls were not receiving sufficient training in country schools to pursue worthwhile homemaking careers, he concluded that pupils of both sexes would benefit from curriculum revisions that provided for the special needs of rural youngsters.

Kelley's scheme for implementing his plans demonstrated ingenuity. He advised school masters and mistresses to "teach children to plant seeds and watch their growth" by placing flowerpots "in every window where the sun shines." "Many a simple lesson in nature is easily taught," he observed, "and it is a singular fact that such lessons impressed upon the mind can never be erased." [9]

Requests for incorporation of agricultural subjects in elementary and secondary school curriculums were on the agenda of every National Grange assembly from 1877 to 1900. Year after year, the body adopted resolutions embodying Kelley's proposals offered in 1877. The first of these came November 29, 1878, and was put forward by Tennessee State Master T. B. Harwell, who recommended that the several states "have the study of the elementary principles of agriculture introduced into the public schools by legislative enactment."

The drive reached its peak in 1897, when Alfred C. True, director of the United States Office of Experiment Stations, gave an address entitled "A Plea for High School Courses in Agriculture," at the Na-

8 National Grange, *Proceedings*, 1877, pp. 41–42.
9 *Ibid.*

tional Grange meeting. True told the delegates that grammar school children were too young to understand complex biological principles, but they could learn to appreciate and to observe plant and animal life. A keener sense of perception could be developed among students, he asserted, by having them plant gardens, observe insects, and engage in other aspects of nature study. At the same time, he recommended that high schools offer specialized instruction in farming practices and in natural sciences, and that teachers of these subjects be given appropriate tutelage in agricultural colleges.[10]

State and local granges also campaigned for curriculum revision. In fact, several movements were already under way when Kelley introduced his ideas in 1877. Tennessee grangers suggested as early as 1875 that school subjects should "embrace more fully our interests as farmers," and a year later Mississippians began their crusade for improved educational programs. Throughout the 1880s and 1890s, aggressive patrons from all across the country struggled to make educators see the need for adjusting classroom work to the toil of the farmyard. Inquisitive farmers did not understand why their children were studying "dead languages" and not receiving specific instruction in any of the disciplines which should make them better tillers of the soil.[11]

The ingenuity displayed by some grange spokesmen while engaged in the serious business of pleading for reassessments of public school curriculums broke up the boredom of repeated drives for the same coveted goal. Their farcical and sometimes humorous anecdotes spiced up many dull meetings and added new life to the movement for better rural educational opportunities. Master S. C. Carr of the Wisconsin State Grange was one of the most skilled storytellers in the organization. He gave audiences samples of his wit almost every time he spoke. In 1883 he woke up the sleepy-eyed portion of his audience with a clever tale which emphasized the need for revising school courses taken by farmers' children. He told a story about a meticulous Bostonian "who parted his hair in the middle and pronounced 'neither' ny-i-ther, who had his kindling wood split according to plans and specifications. The sticks were of the same length and thickness, and tied with black tape in bundles of uniform size, and piled due east

10 *Ibid.*, 1877–1900, general.
11 Tennessee State Grange, *Proceedings,* 1875, p. 51; Mississippi Patrons of Husbandry, *The State Grange and A. & M. College* (n.p., n.d.), 1; Portsmouth *Virginia Granger,* December 14, 1882; Illinois State Grange, *Proceedings,* 1897, p. 46; Wisconsin State Grange, *Proceedings,* 1885, pp. 32–33.

and west in his cellar." But, Carr asked his listeners, "did it burn any better for this fussing?" The speaker then drew a fitting analogy between the New Englander's particularness and the educational approach stressing only the fine points of grammar and arithmetic and ignoring the practical side of knowledge.[12]

Patrons of Husbandry recognized that subject content was not the only factor contributing to a wholesome school experience. Their examinations of the nation's educational systems revealed other startling deficiencies which convinced them that the order's curriculum demands needed to be coupled with requests for teachers to change their teaching methods. Grangers deduced from their investigations that the caliber of instruction and the method of presentation were equally important and that the traditional pedagogic approach was unsatisfactory. For centuries the stimulus-response educational theory had been the basis of formal learning experiences, and students' academic abilities were almost always judged according to their rote-memory capacities. Rarely were allowances made for pupils gifted with exceptional reasoning powers. Although they were not sophisticated enough to understand the full ramifications of their suggestions, grangers by questioning archaic methods were saying the same things that John Dewey and William James were to state several years later when they disturbed the teaching profession with their ideas. After viewing the way rural schools taught their charges, Patrons of Husbandry concluded that children would certainly learn more by doing than by memorizing.[13]

Kansans in the order were particularly perceptive in identifying the faults of rote-memory techniques. In 1884 the state grange committee on education issued a comprehensive paper advocating teaching procedures essentially like those expounded by progressive educationists in the twentieth century. These Kansas committeemen noted "that

[12] Wisconsin State Grange, *Proceedings*, 1883, pp. 13–14.
[13] Gay W. Allen, *William James: A Biography* (New York, 1967); Reginald D. Archambault, *Dewey on Education: Appraisals* (New York, 1966); Melvin C. Baker, *Foundations of John Dewey's Educational Theory* (New York, 1955); Edward C. Moore, *American Pragmatism: Pierce, James, and Dewey* (New York, 1961); George R. Geiger, *John Dewey in Perspective* (New York, 1958); Lawrence A. Cremin, *The Transformation of the School: Progressivism in American Education, 1876–1957* (New York, 1964), 21–22, 41–50; Ann M. Keppel, "Country Schools for Country Children: Backgrounds of the Reform Movement in Rural Elementary Education, 1890–1914" (Ph.D. dissertation, University of Wisconsin, 1960); Kenyon L. Butterfield, "A Significant Factor in Agricultural Education," *Educational Review*, XX (1901), 301–306.

the education in the country schools as now conducted is to a prevailing degree on a plan of memory drill, and of studies involving abstract reasoning, and abstract principles, not suited to the ages and capacities of the children taught in such schools." In lieu of this mode of teaching, the investigators substituted "a plan of object teaching" based on the correct assumption that direct learning experiences were more meaningful than abstract ones. Their detailed report reflected advanced thinking in the field of educational psychology and added prestige to the order's claim of being primarily an educational and social organization.[14]

Complaints against rote-memory teaching were only part of the Grange's campaign for better instruction in rural schools. Illinois grangers were especially concerned about the quality of teaching in their state. They did not think that their children received adequate lessons in spelling and reading to prepare them for adulthood, and they also lamented the way youngsters were rapidly advanced without enough reviewing of essential material. On a more positive note, they felt that all teachers should be at least twenty years old, and they fought for the elimination of all legal barriers to prospective instructors who had not studied Latin.[15]

Grange struggles to upgrade teachers' qualifications were commonplace. The order's drive for better instructors generally followed two directions. Patrons supported demands for increased teachers' salaries, and they pleaded for improved normal-school programs. An 1896 National Grange committee investigating teachers' pay schedules discovered alarming statistics showing the financial plight of the nation's pedagogues. They found that the average schoolmaster received only $46.39 a month, while at the same time, female instructors averaged approximately $8 less. These deplorable salaries disturbed the committee members, and they concluded that more money would have to be allotted for education in order to attract high caliber, professional people. In addition they proposed a plan resembling a modern community college program. Their solution to the critical teacher shortage called for the establishment of departments at "every first-class secondary school for the training of its graduates." [16]

In a few instances grangers displayed a narrow sectionalism in considering teacher qualifications. Members of a subordinate grange lo-

14 Kansas State Grange, *Proceedings*, 1884, pp. 32–33.
15 Illinois State Grange, *Proceedings*, 1892, p. 58; 1897, p. 35; 1898, p. 27.
16 National Grange, *Proceedings*, 1896, pp. 144–45.

cated at Bowling Green, Mississippi, thought their schools would be improved if all northern teachers were expelled from the area. These provincial and bitter souls reasoned "that the harmony, peace, and prosperity of our country demand that our public schools should be conducted by competent teachers of our own people and not by foreigners." [17]

Such questionable views were occasionally rejected by the good sense of members. When State Education Chairman James Oliver of Massachusetts proposed ridiculous theories about teaching at the 1899 session of the state grange, delegates voted to reject his report. Oliver was a medical doctor who liked to dabble in other fields. His varied activities included teaching in public schools, serving as a school-board member, and being a gentleman farmer. Among other things, Oliver attributed youngsters' defective eyesight to teachers' extensive employment of blackboards, and he also complained that instructors were being "hemmed in" and "hampered" by "too much instruction, plans, and methods." Fortunately, the delegates attending the state assembly were not swayed by the physician's diagnosis of the ills of schools. In fact, they refused to endorse Oliver's ideas and offered their own set of more practical resolutions. Moreover, by the next annual assembly, Oliver's office had been taken from him.[18]

Rural schools were not as well financed as urban schools, and this inequity produced complaints. Grangers wanted their offspring to receive the same opportunities as city youngsters. The ire of New Hampshire patrons was aroused in 1890 after an investigation conducted by the state grange committee on education produced pertinent but alarming facts concerning the extent of the discrepancy existing between metropolitan and country school districts in the Granite State. The committee members found that the expenditure range per pupil in cities was from $7.50 to $15.96, while small towns were spending only $4 to $5 for each student. The committeemen concluded from their research that some system of state aid was needed to equalize the funds available for each locality.[19] Washington state grangers discovered that the same situation prevailed in their state. To their astonishment they learned that rural teachers were being paid about $90 for a three-month term and that the remuneration given instructors in first-class cities ranged from $1,000 to

---

[17] Ferguson, "Agrarianism in Mississippi," 175.
[18] Massachusetts State Grange, *Proceedings*, 1899, pp. 59–65; 1900.
[19] New Hampshire State Grange, *Proceedings*, 1890, pp. 90–91.

$1,800 for a ten-month academic year. The state lecturer bitterly questioned whether farmers "were equal before the law, or are we to create the wealth in the first-class cities to rob our children of that education which they are entitled to from the state." [20]

Discovering as they were that quality instruction cost money, grangers did not always object when taxes were increased to provide better schools for their children. Members of the order in Maine adopted a policy statement in 1897 which summed up the feelings of many patrons in regard to this matter. Their resolution stated that they believed in "economy in the expenditure of public money for either national, state, or local purposes," but that their belief in frugality must not be mistaken for "stinginess" in expenditures for education. "Schools are the very life blood of the nation." Grangers showed their firm support for education and their willingness to pay higher taxes for this purpose by developing numerous plans for bringing additional revenue into school coffers. The order consistently called for progressive legislation which reflected the organization's liberal inclinations. Most of the patrons' schemes involved some form of state or federal aid to local school districts.[21]

In the early stages of the Grange's fight to bring equity to the nation's schools, the proposals flowing from state and local assemblies were less sophisticated than those advanced later. To provide more money for their state's public education system in 1875, Alabamians asked for a tax levy against all dogs in their state. Members of two Missouri subordinate chapters felt that a $1 poll tax would solve the fiscal problems connected with the state's schools. Their suggestions were incorporated in petitions forwarded to the state legislature in 1877. At first the order had favored state and national assistance, but the organization still wanted local communities to retain control of their schools. The pendulum, however, was swinging further left. By 1896 the National Grange shifted its position, and now reported that it wanted all schools controlled and financed by "one grand national system." [22]

Besides seeking curriculum reforms and improved instruction,

20 Washington State Grange, *Proceedings*, 1894, p. 22.

21 Maine State Grange, *Proceedings*, 1897, p. 61; National Grange, *Proceedings*, 1883, p. 89; Massachusetts State Grange, *Proceedings*, 1897, p. 67; Texas State Grange, *Proceedings*, 1884, pp. 33–34; New Hampshire State Grange, *Proceedings*, 1898, pp. 117–18.

22 Alabama State Grange, *Proceedings*, 1875, p. 35; Missouri *House Journal* (1877), 606, 855; National Grange, *Proceedings*, 1896, p. 143.

grangers also demanded free textbooks. They claimed first of all that such an arrangement was more economical since the state and national governments could purchase books at quantity rates, while individuals could not. Patrons also maintained that many children from indigent families were discouraged from attending school by the prohibitive burden that book purchases imposed on them. Furthermore, they contended that if schoolbooks were bought less frequently and if government purchases were open to bidding, the publishing "combination" would soon lose its strangling power.[23]

Although numerous civic and educational associations cosponsored with the Grange a free-textbook push, the drive still did not produce the results sought by its advocates. Most state and national lawmakers were too conservative to put the government in the textbook-distribution business. The fate of a free-textbook bill introduced at the 1895 session of the West Virginia House of Delegates was much the same as other proposals of this type. On January 28, Representative Frank H. Smith introduced a bill providing for a system of textbook dissemination, and it was referred to the Committee on Education. While the bill was being considered in committee, the state grange tried to drum up support by forwarding a resolution which was read on the floor of the assembly January 30. Among other things, the order asked for exclusion of "all books of a partisan or sectarian nature" and requested "that the contract [be] for the very best books at the very best prices obtainable." In the end, however, the grange's efforts were futile; the bill was brought from committee without recommendation February 7. Six days later Smith obtained unanimous consent to have his bill withdrawn, and once again the Grange's push to secure a textbook law was foiled by legislative hostility.[24]

While most states did not enact free-textbook legislation, there were a few exceptions. In schoolbook reform, Missouri and California led the way with progressive proposals. In the former state, the legislature adopted a uniform textbook law on March 26, 1874. According to provisions of the statute, presidents of local boards of education and dis-

---

[23] Washington State Grange, *Proceedings*, 1890, p. 14; New Jersey State Grange, *Proceedings*, 1891, p. 56; Wisconsin State Grange, *Proceedings*, 1875, pp. 82–84, 88; Texas State Grange, *Proceedings*, 1884, pp. 41–42; Pennsylvania State Grange, *Proceedings*, 1887, p. 37; Illinois State Grange, *Proceedings*, 1876, p. 36; 1890, p. 67; 1892, p. 39; National Grange, *Proceedings*, 1879, p. 85; 1896, p. 143; Chicago *Prairie Farmer*, December 12, 1874; Greene County (Ill.) Pomona Grange No. 71, Secretary's Minutes, November 13, 1890, Illinois State Historical Society, Springfield.

[24] West Virginia House of Delegates, *Journal* (1895), 196, 295, 460.

trict school directors were scheduled to meet every five years to select manuals of instruction. The law was never utilized and was ultimately repealed. California, on the other hand, adopted a feasible plan which was put into operation. A special state board of education consisting of the governor, the superintendent of public instruction, and the principals of the state's normal schools was created and empowered to choose texts for the state. The superintendent of state printing was to publish and sell books at cost. All selections remained in use for a minimum of four years. Overall, this system saved citizens of the state an average of 68 percent.[25]

Extensions of school terms and stricter enforcement of attendance laws were two other demands presented by granges for the improvement of education in the nineteenth century. Most rural children were in school only three or four months a year, and many youngsters did not attend at all because truancy laws were not rigidly enforced. Studying the situation convinced many grangers that something needed to be done. The order viewed illiteracy as an "evil" and a "menace" to society which had to be stamped out if the republic were to survive. Grangers launched campaigns to extend the academic year from four to nine months and to provide legal means for coercing parents into sending their children to school.[26]

Grangers also were concerned about the moral fiber of the nation, and they thought that it was the school's responsibility to instill certain values in students. At the same time, the order clung tenaciously to the doctrine of separation of church and state. In a policy statement in 1892, the Illinois State Grange stated that morality and virtue should be stressed in schools, whereas "no religious or sectarian dogmas should be taught, and no portion of the public school fund should ever be applied to the support or encouragement of private or sectarian schools." [27]

Paralleling the organization's commitment to morality was its desire to have pupils receive special lessons in Americanism. It was during the late nineteenth century that seeds of super-patriotism were

25 Macon *Missouri Granger*, December 8, 1874; Wisconsin State Grange, *Proceedings*, 1890, p. 7.

26 National Grange, *Proceedings*, 1879, p. 99; 1891, p. 186; Colorado State Grange, *Proceedings*, 1888, pp. 18–19; Wisconsin State Grange, *Proceedings*, 1875, p. 91; New York State Grange, *Proceedings*, 1879, pp. 92–93; Ohio State Grange, *Proceedings*, 1882, p. 6; Washington State Grange, *Proceedings*, 1874, p. 57.

27 Illinois State Grange, *Proceedings*, 1892, p. 61.

taking root in the United States. Tirades of fiery-tongued orators and of journalists became common, influencing many individuals, including several officials of state granges. Many leading members of the order suddenly became inordinately proud of their country's heritage, and the natural result followed. Patrons now wanted to make sure that their children had the same emotional attachments to the United States as they had. Consequently, they commenced campaigns which were prefaced with the same underlying ring of right-wing fanaticism which was to mark mid-twentieth-century agrarianism. In these drives, schools were usually singled out as the ideal agent for imbuing youngsters with sentimentality. The same overall approach was used everywhere, with variations from place to place. Men like George Washington and Thomas Jefferson were now placed on pedestals, and documents like the Declaration of Independence and the Constitution were given positions next to the Bible. Moreover, patrons now demanded that schools ignore any history that did not amount to propaganda.

Mary Sherwood Hinds of the Michigan State Grange was undisputed leader of the grange crusade. In 1897 she distributed two thousand patriotic rituals to Michigan schools, and during the next year Sister Hinds and her co-workers insisted that schools observe all legal and national holidays, have copies of the Declaration of Independence posted conspicuously in every classroom, and teach children to salute daily the flag. Largely because of their work, the state passed a law requiring all schools to fly the Stars and Stripes over every building. The ladies of the Delaware State Grange also were actively engaged in pushing for similar objectives. In 1891 they convinced state school officials to set aside February 22 as Young Patriots Day. On that date teachers were instructed to exhort the virtues of American life and history.[28]

One problem which seldom confronted grangers when they worked for educational improvements was that of solidarity, but nevertheless when divisions of thought did occur, intense rivalry and bitterness usually followed. One issue which tended to produce the most resentment and which disrupted chapters was school consolidation. Although many patrons thought that their children would benefit from graded elementary schools, others opposed centralization because they

[28] Fred Trump, *The Grange in Michigan* (Grand Rapids, 1963), 59, 63; Delaware State Grange, *Proceedings*, 1890, p. 34; 1891, p. 21.

did not want their offspring traveling long distances to and from school. Moreover, they did not want district schools closed since they offered unexcelled opportunities for social intercourse.

Chairman James Oliver of Massachusetts feared the results that would accrue from abandonment of local schools. He told delegates attending the 1899 session of the state grange that pupils did not receive as much personal attention in consolidated schools because "instruction fired at the head of one must hit the heads of all and produce the same effect." He also maintained that the "iron-clad system of grading" employed by graded schools was responsible for the high dropout rate.[29]

Opponents of school consolidation effectively blocked most proposals calling for graded schools, but there were some noteworthy exceptions. Oliver's emotional harangues, for instance, were not enough to offset the wisdom and sound arguments presented for centralization, and his plan for turning back the clock and restoring district schools was rejected. After hedging on the question of school consolidation for a few years, proponents finally garnered enough support at the 1900 meeting of the Illinois State Grange to pass a resolution favoring graded township schools.[30]

Advocates of school consolidation had no warmer friend than Sister Mary E. Lee of Fairfield, Ohio, whose appeals were carried by the *Ohio Farmer*. In 1897 she urged patrons to "push along the work of centralization of schools" because "country schools can not have the advantages of libraries, maps, etc. A country school is fortunate if it gets a dictionary." Moreover, a beneficial exchange would take place since merger "will break down the barriers between town and country.... The country child will acquire grace of manner, ease in speech, and quickness of thought." On the other hand, "the town child will learn lessons of endurance, self-denial, and integrity from his sturdy country cousin." A few weeks later the state grange surveyed schools in three counties in which consolidation had taken place and reported that instruction had improved and educational costs had decreased. The combination of Mrs. Lee's writings and the committee's

29 Massachusetts State Grange, *Proceedings*, 1899, p. 61.
30 New York State Grange, *Proceedings*, 1890, pp. 119–20; 1892, p. 131; Trump, *The Grange in Michigan*, 59; Michigan State Board of Agriculture, *Annual Report*, 1897–98 (Lansing, 1899), 671; Massachusetts State Grange, *Proceedings*, 1899, pp. 64–65; Illinois State Grange, *Proceedings*, 1898, p. 27; 1900, p. 35.

findings convinced Ohio patrons to endorse the state's centralization program.[31]

One of the main advantages of school consolidation was development of worthwhile high school programs, something desired by the Grange. In fact, patrons began urging state legislatures to make universal opportunities for secondary education available to all children as early as 1875 when the Wisconsin State Grange accepted proposals offered by State Superintendent of Public Instruction Edward Searing and by a three-man board established by the Wisconsin State Teachers' Association. Both Searing and the teachers had asked the order to join them in a campaign for the establishment of a system of intermediate schools, "easily accessible to all, offering to those who have finished the elementary course of the primaries a sound supplementary instruction in the ordinary academic studies, and especially in the natural sciences in their application to mechanics, manufactures, and agriculture." Searing and the teachers' group wanted the state to adopt the Maine Plan for Secondary Schools. According to this scheme, one or more towns were given authority to establish and maintain high schools, with revenues coming from proceeds of a common tax and from private contributions. This system had already been tested successfully in Maine for two years, when the Wisconsinites began pleading for its adoption in their state.[32]

The National Grange gradually unfolded a comprehensive but not always consistent program to bring "city advantages for country children," the order's "rallying cry" to equalize the nation's educational opportunities and to make high schools easily accessible to all youngsters. Rural boys and girls rarely had the same opportunities for secondary educations as city children. Chairwoman Sarah Baird of the national committee on education recognized this shortcoming in 1896. At the time, only Massachusetts had a state law requiring high school opportunities for every child in the state. Nowhere else, she noted, did grammar school graduates always have convenient access to intermediate institutions. In the country, teenagers who wanted to continue their educations often had to board in town, a procedure which few rural families could afford, or they had to step directly into college, a difficult transition that placed these students at a competitive disad-

---

[31] Cleveland *Ohio Farmer*, November 18, 1897; Ohio State Grange, *Proceedings*, 1897, p. 32.
[32] Wisconsin State Grange, *Proceedings*, 1875, pp. 81–82, 85–88.

vantage. What was needed, claimed Mrs. Baird, were additional secondary schools located near enough to all prospective students.[33]

By their very nature city schools offered certain advantages which were unobtainable in rural institutions. The latter's small enrollments made it almost impossible for them to compete with large urban high schools. By drawing large numbers of students, metropolitan institutions could afford to provide better facilities, such as adequate libraries and gymnasiums. Although most grange leaders knew that country districts were handicapped, they refused to abandon their efforts to find means of closing the gap. Their refusal to capitulate helped to uplift educational offerings of many rural schools in the late 1800s.

Advent of a state-operated circulating library program in Michigan was one result of granger determination. In 1895 the state librarian solicited assistance of the state grange in his campaign to secure an adequate appropriation for this project. Grangers in turn responded enthusiastically, and the librarian and his agrarian colleagues effectively coalesced their demands into one movement for establishment of a mobile lending service. The state legislature, seeing merit in the plan, passed a revenue act earmarking funds for the program. In the years that followed, the order worked vigorously for expansion of initial services. Thus, the Grange can be given partial credit for providing Michigan's rural residents with an ample supply of reading materials.[34]

Not only did grangers seek to gain the same educational opportunities for their offspring as were available to city youngsters, but they also wanted to give women an equal voice in the administration of schools. Women were legally barred from serving as school board members, school inspectors, and county and township superintendents of education. Laws setting sex qualifications for voting and holding office were constant sources of irritation to the reform-minded Patrons of Husbandry. In 1875 delegates attending the Wisconsin State Grange showed that they were sympathetic with the feminist cause by joining State Superintendent Searing in support of legislation calling for the elimination of these forms of discrimination.[35]

The order also sought better relationships between the teacher and the farmer. The Hesperia Movement in Michigan was the most outstanding example of these efforts. According to its plan, educators

33 National Grange, *Proceedings*, 1896, p. 142.
34 Trump, *The Grange in Michigan*, 56.
35 Wisconsin State Grange, *Proceedings*, 1875, pp. 84, 88.

united with grangers to form parent-teacher associations. Through cooperation, participants became acquainted with each other's activities. Joint meetings proved to be so useful that in 1897 the Michigan State Grange requested all of its subordinate chapters to form similar associations in their counties.[36]

Teachers and school administrators willingly fraternized with grangers because they found the sessions of local chapters to be very convenient forums for reaching their pupils' parents. When asked to comment upon the merits of the Grange, one anonymous county school commissioner said he did not know how he would have been able to contact "the parents of the children if it were not for the grange. It is the only place where I can be sure of getting together with them in order to discuss the plans upon which I feel I must have their opinion and co-operation." [37]

The order encouraged its members to visit schools and to become acquainted with their operation. By being interested in activities carried on by these institutions, parents gave encouragement to both students and teachers "to do their best" by showing that the community was concerned about its educational system. At the same time, the Grange found that periodic visits kept school officials on their toes and eliminated waste. Master William G. Wayne of the New York State Grange told delegates attending the 1879 conclave that he discovered that many district school commissioners were not performing as well as they should. As a safeguard he suggested that subordinate chapters take it upon themselves to guarantee that these officials carry out their delegated duties by making surprise visits to neighborhood schools.[38]

Whether grange pressure caused any wave of change or amelioration is debatable, but one thing is certain—the order definitely laid the groundwork for future programs in schools. These included vocational agriculture, "progressive" teaching methods, agricultural education training in pedagogics departments, strict enforcement of compulsory school attendance laws, school year extension, rural high schools for all children, and curriculum adjustment to meet the needs of the community. In addition, farmers became increasingly more acquainted with the public educational system, its programs, and its future.

[36] Trump, *The Grange in Michigan*, 49–50, 66.

[37] Jennie Buell, "The Educational Value of the Grange," *Business America*, XIII (January, 1913), 54.

[38] Maine State Grange, *Proceedings*, 1897, p. 47; New York State Grange, *Proceedings*, 1879, pp. 11–12.

CHAPTER IV

The Grange and Higher Education

In 1862 Congress followed advice of Jonathan B. Turner and other agrarian spokesmen and passed the Morrill Land-Grant Act. After Abraham Lincoln signed the bill into law, every state received thirty thousand acres of public land or an equivalent amount of land scrip for each representative in Congress. Proceeds coming from the sale of the grant were to be used for education in military science, engineering, and agriculture. The types of colleges designated by state legislatures to execute the land-grant program varied from state to state. But generally schools fell into three major categories: existing state and private universities such as the University of Minnesota and Yale University, established agricultural colleges like Michigan State University, and newly created institutions like Purdue.[1]

Although the National Grange first expressed an interest in land-grant colleges in 1874, it was not until 1876 that the national body suspected that terms of the Morrill Act were being violated. Instead of employing money from the sale of allotted lands for the designated purposes, grange leaders now feared that some universities were using their funds for teaching liberal arts. Reacting to its suspicions, the 1876 national delegation created a special standing committee to look into the matter. Its members were instructed "to inquire what colleges have been established under the said act, what donations have been made to said colleges other than the donations of Congress, and what success they have attained in the prosecution of the work proposed to them in the law creating them." They were also authorized "to look over the whole ground of agricultural education and to report to this body ... what has been done and what ought to be done." Further-

---

1 U.S. *Statutes at Large*, XII, 503–505; Eugene Davenport, "History of Collegiate Education in Agriculture," Society for the Promotion of Agricultural Sciences, *Proceedings*, 1907 (Lansing, 1907), 43–53. For a full account of the Morrill Act and its immediate application, see Allan Nevins, *The State Universities and Democracy* (Urbana, Ill., 1962) and Earle D. Ross, *Democracy's College: The Land-Grant Movement in the Formative Stage* (Ames, Iowa, 1942).

more, the parent organization requested its state chapters to create similar permanent groups to investigate conditions in their respective jurisdictions.[2] At the same time, delegates took a firm position concerning the arrangement in those states where schools of agriculture endowed with land-grant funds had been grafted to an existing university. Patrons resolved "that these colleges [the land-grant institutions] ought to be ... separate and distinct schools, where science, as applied to agriculture, may be taught to farmers' children, fitting them for the high calling of farmers."

In the years that followed the National Grange continually sought to separate colleges of agriculture from universities. On February 29, 1884, the cause won a limited victory when Representative D. Wyatt Aiken's resolution asking for a congressional investigation into usage of Morrill Act funds was approved. But when Congress took no further action after the Committee on Education made an inquiry, grangers resumed their fight.[3]

These efforts reached a peak in 1892, when the order's committee on education presented to the Grange a thorough report on the status of agricultural education in land-grant institutions. It found that in the thirteen states where existing colleges had been assigned the task of implementing the purposes of the Morrill Act, there were only 770 students pursuing agriculture, and only 304 had been graduated with degrees in agriculture. On the other hand, in thirteen states where separate schools existed, there were 4,386 students and 2,616 graduates. From these records, committeemen deduced that "none of the Agricultural Colleges which are connected with classical institutions have been successful in imparting agricultural education, and a portion of them have been dismal failures, while ... the independent Agricultural and Mechanical Colleges are, without exceptions, eminently successful." [4]

After examining the results of its investigation, the national committee on education became more convinced than ever that Congress should pass a law severing ties between agricultural and mechanical colleges and "classical" institutions. The delegates adopted the committee's resolution to this effect; they also drafted a petition embodying these principles and sent it to congressmen. On June 6, 1892, Sen-

---

2 National Grange, *Proceedings*, 1874, p. 58; 1876, pp. 106–108.

3 *Ibid.*, 1882, p. 108; 1883, p. 120; 1889, p. 97; 1891, pp. 180–81; 1893, p. 148; United States *Congressional Record*, 48th Cong., 1st Sess., 1496–1497.

4 National Grange, *Proceedings*, 1892, pp. 90–100.

ator James H. Kyle of South Dakota presented their entreaty "praying for the separation of agricultural and classical colleges, and for the establishment of separate agricultural and mechanical colleges in all cases where Government aid is granted." Two days later, Senator Daniel W. Voorhees of Indiana presented a similar demand.[5] Congress was unimpressed, but alarmed administrators began revising their colleges' offerings to placate their critics.

Even before the parent organization became involved in the land-grant college controversy, several state granges had tried to create private agricultural institutions. The Mississippi State Grange began considering a proposal for establishing such a school in 1873, but the scheme was left pending until 1875 when the committee on education, after studying the plans' merits, rejected them because of a lack of funds in the state grange treasury. The proposition, however, was not completely forgotten. At the next session of the Mississippi grange, a special five-man commission reevaluated the whole question and suggested in its report issued December 15, 1876, that the state allocate revenues arising from the land-grant fund to the order for the purpose of developing a semiprivate agricultural and mechanical college.[6]

West Virginia and Texas grangers also flirted with the idea of establishing private agricultural colleges. Patrons in the Lone Star State considered purchasing an experimental farm where rural youths could learn the latest methods in scientific agriculture, but the state body discarded the scheme because of insufficient funds and because of the establishment of Texas Agricultural and Mechanical College in 1876. In West Virginia, members of the order came closer to seeing their plans instituted. In January, 1875, the state legislature chartered Jefferson County Agricultural College, a school planned by grangers. The institution never opened, however, because its founders could not secure a public land grant. The final setback came when Daniel B. Lucas, one of the original supporters and a granger, was appointed to the West Virginia University Board of Regents, and he promptly switched his allegiances.[7]

[5] *Ibid.*, 101; United States *Congressional Record*, 52nd Cong., 1st Sess., 5050, 5130.

[6] Ferguson, "Agrarianism in Mississippi," 182; Mississippi State Grange, *Proceedings*, 1875, p. 16; Mississippi State Grange, *The State Grange and A. & M. College* (n.p., n.d.), 1.

[7] Robert L. Hunt, *A History of Farmer Movements in the Southwest, 1873–1925* (College Station, Tex., 1935), 11–12; National Grange, *Proceedings*, 1877, pp. 128–29; William D. Barns, "The Influence of the West Virginia Grange upon Public

As these private college plans were being cast aside, the Mississippi State Grange began its campaign to strip the state university at Oxford of its land-grant status. The first suggestion came in 1874 at the annual state conclave of the society, but need for such action did not become imperative until 1876 when the University of Mississippi completely abandoned its agricultural program.[8]

When the state grange convened at Jackson on December 12, 1876, the university's action received immediate attention. J. B. Yellowley of Madison County recognized full well the value of education to agriculture, and he was greatly disturbed by the university's abandonment of its agricultural program. He quickly offered a resolution creating a special five-man committee to consider the practicality of establishing a new college to replace the existing institution as the official recipient of the land-grant fund. The delegates approved Yellowley's proposal and set up an investigating team to determine what could be done. Four days later the committee members made their report. They advised the state grange to establish another committee to develop plans for "a purely Agricultural and Mechanical College" and to submit the scheme to the state legislature.[9]

When the Mississippi General Assembly met in 1877, the legislators took up state grange requests for a transfer of land-scrip receipts to another institution. Members of the special grange committee studying agricultural education had recommended chartering of a semiprivate college operated by their order but subsidized by funds arising from the Morrill Act. But J. B. Yellowley, now a state representative and the grange's spokesman in the legislature, had something else in mind when he wrote a bill dealing with the issue. Yellowley called for establishment of a new public institution. The proposed piece of legislation did not conform to the grange's initial specifications, but grangers still supported it. They flooded the legislature with petitions praying for its adoption, and friendly lawmakers read many of their remonstrances from the floor of the assembly. These efforts, however,

Agricultural Education of College Grade, 1873–1914," *West Virginia History*, IX (January, 1948), 130–33; J. Burns Huyett, "Early Grange Activities in Jefferson County [West Virginia]," Jefferson County Historical Society, *Magazine*, V (December, 1939), 6.

8 John K. Bettersworth, *People's College: A History of Mississippi State* (University, Ala., 1953), 18; Ferguson, "Agrarianism in Mississippi," 181.

9 Mississippi State Grange, *The State Grange and A. & M. College*, 1; J. M. White, "Origin and Location of the Mississippi A. & M. College," Mississippi Historical Society, *Publications*, III (1900), 346.

were not enough to move the state's lawmakers. Judging from their actions, they saw no need for another state college; they rejected Yellowley's bill.[10]

The rebuff did not discourage Mississippi patrons, instead, it made them more determined. On the first day of the 1877 state grange meeting held at Holly Springs, John Robertson of De Soto County resolved "that the Legislature of the State shall establish an Agricultural College in accordance with the intention of the act of Congress . . . and that no further delay nor frittering away of the [land-grant] fund will be quietly tolerated by the agriculturists of the State." [11]

At the next session of the state legislature, William F. Tucker of Chickasaw County submitted a bill which was essentially the same as the one presented at the last session. In conjunction with the presentation of Tucker's bill there was a massive grange campaign for bringing pressure to bear upon lawmakers. Patrons sent countless memorials to their representatives. Their unyielding efforts succeeded in arousing the solons; the agricultural college bill cleared its final legislative hurdle February 22, 1878, and it became law six days later. Thus, grangers had received what they wanted—a college where their sons could study latest farming techniques. In large measure, it was their determination and influence that was responsible for formation of Mississippi Agricultural and Mechanical College, now Mississippi State University.[12]

Mississippi was not the only state below Mason and Dixon's line where weight of grange influence affected development of land-grant institutions; separatist activities also abounded in other areas of the South. North Carolina grangers' demands were very similar to those pushed by their Mississippi brethren. In 1876 North Carolinians began complaining that President Kemp Plummer Battle's token attempt to provide agricultural instruction at the University of North Carolina was not paying dividends. But the state legislature waited until 1887 to act upon the state grange's demands. In that year, Representative Augustus Leazer and Senator Sydenham B. Alexander, a former state grange master, introduced a bill stripping the university

10 Mississippi *House Journal* (1877), 185; Ferguson "Agrarianism in Mississippi," 182; Bettersworth, *People's College*, 18–19; White, "Origin and Location of the Mississippi A. & M. College," 346.

11 Mississippi State Grange, *The State Grange and A. & M. College*, 1.

12 Mississippi *House Journal* (1878), 42, 136, 189; Mississippi *Senate Journal* (1878), 40–41, 409–10; Starkville (Miss.) *Southern Live-Stock Journal*, June 2, August 4, 1881.

of its land-grant privileges and establishing a new state agricultural and engineering college. Their bills prompted a bitter fight between advocates of separation and friends of the Chapel Hill institution. The latter ultimately lost, and the Leazer-Alexander bill chartering North Carolina State College at Raleigh became law. Although the role of Patrons of Husbandry in North Carolina was not as clear-cut as it had been in Mississippi, the Carolina patrons could still claim some credit for the founding of the state agricultural college.[13]

Elsewhere in the South and in border states, there were rumblings of discontent. These were occurring in Kentucky, South Carolina, Louisiana, and West Virginia.[14] Grangers in the latter state became upset with their land-grant arrangement after Overseer Thomas C. Atkeson had told them at the 1891 session that West Virginia had only a "so-called" agricultural college and experiment station at Morgantown.

Delegates accepted Atkeson's assessment without question; to them his experience in agricultural education qualified him to evaluate facilities and instruction available at West Virginia University. He had studied agriculture at the University of Kentucky and was known to be a keen student of scientific farming practices. In view of the fact that Atkeson thought the institution was "an entire failure" and should lose its land-grant status, it was no surprise that the state grange adopted his recommendation calling for detachment of agricultural responsibilities from the Morgantown school.[15]

The land-grant dispute in West Virginia was temporarily settled by a clever compromise between the Board of Regents and Atkeson. After the legislature had refused to divorce agricultural responsibilities from West Virginia University, the school's trustees soothed the patrons' wounds by creating a chair of agriculture and offering the professorship to Atkeson. The proffered nomination came as a complete shock to Atkeson, but he concluded that it was his moral obligation to accept. He began his duties as the first "professor of agriculture" at West Virginia University during the fall term of 1891.[16]

Atkeson performed well, but ran into trouble with the president

[13] Lindsey O. Armstrong, "The Development of Agricultural Education in North Carolina" (M.S. thesis, North Carolina State College, 1932), 24, 32–34, 39, 61.

[14] National Grange, *Proceedings*, 1892, pp. 93–95, 97–99; Walter L. Fleming, *Louisiana State University, 1860–1896* (Baton Rouge, 1936), 430–31.

[15] West Virginia State Grange, *Proceedings*, 1891, pp. 15–16.

[16] Barns, "The Influence of the West Virginia Grange upon Public Agricultural Education of College Grade," 136–37.

and the faculty, who felt that his courses were "beneath the dignity of so important [an] institution" and "an undesirable invasion into the holy precincts of classical knowledge." Their reservations resulted in a general reorganization of the college staff at the end of Atkeson's second year at the West Virginia institution. He was dismissed, and his tasks were assigned to the horticulturist of the experiment station. This arrangement failed, partly because the horticulturist was shackled with so many responsibilities that he could not do any of them well. In any event, the school of agriculture decayed.[17]

Leaders of the state grange gradually sensed that something was seriously wrong at their university. They became so displeased by the turn of events that they once again pleaded with the legislature to create an independent college of agriculture. Patrons justifiably felt that they had been betrayed, and they did not want any repetition. Consequently, at the 1897 conclave they voted for the erection of a new agricultural school "apart from our State University." Lawmakers ignored that demand, but in 1897 they provided for a complete reorganization of the Board of Regents. Governor George W. Atkinson used power vested in him by the legislature to name an entirely new board. Among the new members was Thomas C. Atkeson, the deposed professor of agriculture and the governor's cousin. Thus, the grange achieved one of its primary objectives; a farmer was now on the governing board of West Virginia University.[18]

Atkeson made the most of his opportunity. He convinced his fellow regents that "the best interests of the University and the State would be promoted by the retirement" of the university president and the removal of the director of the agricultural experiment station; both men were enemies of the Patrons of Husbandry. After these adversaries had been replaced, the regents implemented Atkeson's plan for a College of Agriculture, headed by a dean and having the same status as other colleges of the university. At the same time, regents provided for four-year and two-year courses of instruction in scientific farming and named Atkeson dean of the College of Agriculture. His appointment won the grangers' praise and ended their criticism of West Virginia University. The College of Agriculture was now in the capable

---

17 *Ibid.*, 137–40.
18 West Virginia State Grange, *Proceedings*, 1897, in Charleston *West Virginia Farm Reporter*, February, 1897; Barns, "The Influence of the West Virginia Grange upon Public Agricultural Education of College Grade," 140–43.

hands of a friend and a fellow Patron of Husbandry, and members of the order were satisfied.[19]

The problem of forcing land-grant recipients to use endowments for purposes designated in the Morrill Act was not limited to the South and border states. Several state granges in the North experienced the same difficulties and the same dissatisfaction with their schools. To their dismay they discovered that their colleges neglected agricultural studies and spent their land-grant revenues for disciplines and purposes not covered by the 1862 law. Procedures employed by northern grangers in dealing with colleges suspected of misapplying Morrill funds were basically the same as those used by patrons below Mason and Dixon's line.

Professor William A. Henry of the University of Wisconsin alerted grangers of his state when at a meeting in 1882 he related difficulties he confronted while trying to perform his duties as dean of the state college of agriculture. Henry's memorable address contained a brief synopsis of the Morrill Law, a summary of its application, a detailed report comparing results in Wisconsin with those elsewhere, and a passionate plea for farmers to interest themselves more in the state's collegiate program in agricultural education. Although Henry did not directly ask the patrons to work for separation of the college of agriculture from the university, he implied that such an arrangement would certainly be desirable and would surely upgrade the land-grant program in Wisconsin. In an era when tenure virtually did not exist, professors had to be cautious when they criticized their institutions. Henry, being a practical man, did not dare risk his job by openly pleading for separation. Instead, he relied upon statistics, knowing that figures would carry his message as well as a direct statement.[20]

For instance, he compared states where no separation had taken place with those having independent land-grant institutions and found that the latter cases were dramatically more successful.

Henry's use of statistics showing that Wisconsin trailed in the field of agricultural education hit the state grange like a bombshell. The professor thus provided a spark that generated an order drive for separating the college of agriculture from the university. Admittedly, such prominent patrons as State Master S. C. Carr had already been committed to separation of the institutions before Henry delivered

[19] *Ibid.*, 144–46.
[20] Wisconsin State Grange, *Proceedings*, 1882, pp. 32–37.

Table 11

STATES WITHOUT
SEPARATE AGRICULTURAL COLLEGES

	State College	Enroll-ment	Ag. Students	Mech. Students
Ark.	Arkansas Industrial U.	300	0	12
Calif.	U. of California	224	13	6
S. C.	South Carolina C.	152	6	3
Ill.	Illinois Industrial U.	352	21	41
Ohio	U. of Ohio	330	30	50
Nebr.	Nebraska State U.	284	26	0
N. Y.	Cornell U.	380	24	37
R. I.	Brown U.	270	0	36
Wis.	U. of Wisconsin	367	7	15
Total		2,659	127	200

Table 12

STATES WITH INDEPENDENT
AGRICULTURAL COLLEGES

Colo.	Colorado A. & M.	55
Mass.	Massachusetts A. & M.	98
Miss.	Mississippi A. & M.	275
Kans.	Kansas A. & M.	292
Mich.	Michigan A. & M.	216
Total		936[21]

his remarks. But Carr's suggestions did not have as much impact on grangers as the dean's subtle call to action.[22]

Wisconsin grangers collaborated with Dean Henry throughout the decade by mounting annual campaigns against the state university. The farmers felt their actions against the Madison institution necessary and justifiable because they were being "elbowed out [there] by other professions." [23] Wisconsin patrons were convinced that their children would become victims of class warfare and ridicule if they

21 *Ibid.*, 36–37.
22 *Ibid.*, 16–17.
23 *Ibid*, 1882, pp. 15, 51–52; 1883, pp. 26–27; 1885, p. 24; 1886, pp. 4–5; 1890, pp. 74–77.

attended a mixed university. Most farmers foresaw an inevitable clash pitting agrarian and nonagrarian classes against one another every time the two groups confronted each other.

Leaders of the order played on this fear. Instead of letting it subside, they used it to work members into a frenzy over the question of separating the college of agriculture from the university. This obviously was a desired objective in 1883 when a special committee on the agricultural college made its report. Committeemen predicted "that whoever can join harmoniously incompatible tempers or incongruous materials, can beat nature's laws and make ice and fire embrace each other to the satisfaction of both." They, of course, were referring to the possibilities of merging "an agricultural and mechanical college ... with a college of letters" and making the arrangement operate without friction and animosity.[24]

The grange drive reached a climax in Wisconsin in 1883, when the state agricultural society sponsored a spirited debate on the question of taking away the land-grant privileges from the University of Wisconsin. Aaron Broughton and Dean Henry presented the patrons' case against the Madison school's handling of Morrill funds, while President John Bascom defended his institution. The administrator argued that his university stressed quality, not quantity. At the same time, he conceded that establishment of another college would no doubt increase overall enrollment because "by virtue of locality, it would gather in one hundred students at least, and if you were to cut that agricultural institution in two and put one part in one place and another part in another you would still get more, and if you divided it into four institutions, by virtue of this local force you would still get more in numbers." Bascom then asked his audience if "these four agricultural institutions scattered over the state" would compare favorably to one well financed and centrally located university.[25]

As things turned out, Bascom's stand prevailed. The university was kept intact, and all the grange could claim from its efforts was Board of Regent's acceptance of the Hitt-Vilas Compromise. That agreement provided for a short course in agriculture at the University of Wisconsin with low entrance requirements. Any boy with a grammar school education was now eligible to receive instruction at the college of agriculture for two twelve-week terms spaced over a two-year span. The short course was the only concession the university was willing to

24 *Ibid.*, 1883, pp. 51–52.
25 Wisconsin State Agricultural Society, *Transactions,* 1883, pp. 181–230.

accept, but it was not enough to placate stubborn men like Carr. They continued making their pleas for separation until 1890. Their persistence, however, failed to produce any tangible results. The Grange was rapidly losing its influence in Wisconsin politics, and politicians knew that they no longer had to heed the warnings of the order's spokesmen.[26]

Grangers in other northern states also expressed dissatisfaction with the way land-grant funds were being used in their states. The separatist activities in Minnesota, Vermont, and Connecticut were the most notable. In each case the order succeeded in bringing about changes. In the latter state, Yale University's Sheffield Scientific School had been sole recipient of Morrill money for more than twenty years before the state grange persuaded the legislature to channel all additional revenue accruing from sales of land scrip to Storrs Agricultural College.[27]

The Minnesota State Grange's campaign against the state university yielded a unique revision of the institution's supervision of its land-grant responsibilities. The school kept its status, but the college of agriculture was moved to St. Anthony Park, a community a few miles from the main campus in Minneapolis. Thus for all practical purposes, the college of agriculture was independent. Its students boarded and attended classes at the site, while distance prevented two rival student bodies from polluting one another, a common concern of patrons in the Gopher State.[28]

Farmers generally were of the opinion that their children would be ridiculed by other students for having agricultural interests. In many instances, unfortunately, these fears and suspicions were not without some substance. The *Illini*, student newspaper of the University of Illinois and periodic friend to the College of Agriculture, highlighted the problem in 1890 by carrying the following hypothetical conversation:

[26] Merle Curti and Vernon Carstensen, *The University of Wisconsin*, 2 vols. (Madison, 1949), I, 470–72; W. H. Glover, *Farm and College: The College of Agriculture of the University of Wisconsin* (Madison, 1952), 99; W. H. Glover, "The Agricultural College Crisis of 1885," *Wisconsin Magazine of History*, XXXII (September, 1948), 20, 22–23.

[27] Walter Stemmons, *Connecticut Agricultural College—A History* (Storrs, 1931), 56–57, 64–65; Russell H. Chittenden, *History of the Sheffield Scientific School of Yale University, 1846–1922*, 2 Vols. (New Haven, 1928), I, 228–29.

[28] Andrew Boss, *The Early History and Background of the School of Agriculture at University Farm, St. Paul* (n.p., 1941), 32–37; C. R. Barnes, "The Department of Agriculture," in *Forty Years of the University of Minnesota*, ed. E. Bird Johnson (Minneapolis, 1910), 117–23.

Mr. Bullywag—"How de do."
New Student—"How are you?"
Mr. Bullywag—"New student, eh?"
New Student—"Yes sir, I have just entered the University and think I
shall like it."
Mr. Bullywag—"That so? What course you takin'?"
New Student—"I expect to take the agricultural course."
Mr. Bullywag—(with a smile of heavenly superiority) "Oh! You takin'
agriculture? I'm takin' engineering." [29]

After waiting patiently for the University of Vermont to perform
its duties in the field of agricultural education with some degree of
conformity to the purposes of the Morrill Law, Vermont grangers be-
latedly joined members from other states in seeking a chartering of a
separate farmers' college. In their arguments, patrons occasionally dis-
torted facts and exaggerated the need for detachment.

On November 10, 1890, the prograge Rutland *Herald* moaned
that the institution's experimental farm consisted of "a cow, pig, and
horse, or something of the sort, a farm that is good for nothing." J. L.
Hills, an experiment station chemist at the university when the controversy raged, challenged the validity of this claim. According to
him, "the animal content of the farm was that of the average farm
[being composed of] a dozen or more cows, horses, pigs etc." On another occasion when the editor of the same paper charged that the
legislature had not enacted any of the bills providing for detachment
because of bribery involving the university and the lawmakers, Hills
doubted whether "the univ[ersity] administration was in a position
to pay for votes" and he also questioned whether college stewards
would have been capable of paying "for votes . . . or would have
countenanced such [a] procedure." Regardless of the merit of the
charges, the university still had to upgrade its department of agriculture in order to retain its Morrill program, so the grange was successful in obtaining some valuable concessions from the Burlington
school, including a separate board of trustees for the college of
agriculture. [30]

State grange leaders were never so naïve as to think separation of
colleges answered all their prayers for better farmer education pro-

[29] Richard G. Moores, *Fields of Rich Toil: The Development of the Illinois College of Agriculture* (Urbana, 1970), 58–60.
[30] Guy B. Horton, *History of the Grange in Vermont* (Montpelier, 1926), 36–47;
Vermont State Grange, *Proceedings*, 1891, pp. 15–16; J. L. Hills's Marginal Comments, in Copy 2 of the University of Vermont's Guy B. Horton, *History of the
Grange in Vermont* (Montpelier, 1926), 42–43, 45.

grams. They recognized that decay could set in at detached agricultural schools as quickly as improvement if certain precautions were not taken to insure fulfillment of these institutions' Morrill Law obligations. It was due to this fear of backsliding and because of a corresponding desire to see additional advances made in the field of agricultural education that many state granges created special inspection committees. The latter's close contacts with colleges of agriculture kept other members abreast of changes occurring at these schools and placed the order in a more advantageous position to evaluate strengths and weaknesses of the institutions' land-grant programs.[31]

Wise administrators placated Patrons of Husbandry by making them feel important. They never hesitated to welcome grangers to their campuses and they always treated their guests as cordially as possible because they knew the least reluctancy on their part might be misinterpreted by suspicious farmers. The red-carpet treatment accorded grangers by the president of Purdue University in 1888 was a good example of how to develop a healthy relationship. The college official surrendered his house and resided in a hotel while a state grange committee touring the college boarded in his mansion. In addition, he placed a carriage and driver at his guests' disposal. These gestures were not soon forgotten, for at the next state grange meeting, appreciative committeemen issued a very favorable report glowing with praise for the college, its program, and its administration.[32]

From the outset, Mississippi Agricultural and Mechanical College enjoyed cordial relations with the state grange. When the school officially opened in 1880, Master Putnam Darden was present as a special invited guest of the board of trustees. On numerous other occasions throughout the decade, President Stephen D. Lee and the trustees urged Darden and his fellow grangers to visit the institution and observe its operation.[33]

There were a few instances when grangers turned on their hosts. In 1887 and 1888, special delegations of Oregon patrons deduced from

[31] Vermont State Grange, *Proceedings*, 1891, pp. 51–53; Kansas State Grange, *Proceedings*, 1876, p. 27; Oregon State Grange, *Proceedings*, 1881, p. 12; 1882, pp. 10–11.

[32] Michigan State Grange, *Proceedings*, 1876, pp. 23, 54; New Hampshire State Grange, *Proceedings*, 1890, pp. 84–86; 1891, p. 92; Indiana State Grange, *Proceedings*, 1888, pp. 20–23.

[33] Mississippi State University, Minutes of the Board of Trustees of Mississippi Agricultural and Mechanical College, September 20, 1880, March 28, 1882, Mississippi State University Library, State University; Mississippi State Grange, *The State Grange and A. & M. College*, 4–7.

their visits to the state agricultural college at Corvallis that the state's farmers had been "deceived and outrageously humbugged" by the administration. According to their assessments, instruction was too theory oriented to be of practical value to farmers and the site of the institution was totally unfit for an agricultural experiment station. In their reports, committeemen who had inspected the campus and its facilities made two recommendations. They asked the state grange to withdraw its support from the college, and they advised the body to demand a relocation of the college.[34]

Occasionally, college campuses were sites for state grange gatherings. In most instances colleges themselves were responsible for having their ivy-covered buildings turned into temporary grange halls because they openly coveted the honor of hosting these assemblages. On the other hand, grangers liked to receive invitations because they permitted the organization to save on auditorium rentals and they gave members an opportunity to visit their land-grant college.[35]

Thus, when trustees of Purdue University unanimously voted in 1877 to offer grangers use of the chapel and military hall for their next meeting, patrons wasted no time in accepting the proposal. As with most conventions of this type, the session began with a welcome by college officials. President E. E. White greeted those present: "The similarity of the aims of the university which I represent, and the Order which I address, makes it a pleasant duty . . . to welcome you." During the course of the conclave, members were shown around the grounds and taken into buildings, thereby enabling them to view livestock, peek into classrooms and laboratories, and to their amazement watch an electric lightbulb glow.[36]

Collegiate interests of state granges in the late nineteenth century may be compared to activities of an energetic alumni body. This was especially true of states where land-grant programs were functioning well. In such instances, enthusiasm exhibited by members of the order for their colleges of agriculture was certainly equal to interest shown by most graduates for their alma maters. Across the nation, grangers demonstrated their support for their institutions by urging legislatures to appropriate additional money to these schools. A special Texas State Grange committee investigating needs of Texas A. & M.

34 Oregon State Grange, *Proceedings*, 1887, p. 35; 1888, pp. 48–49.
35 New York State Grange, *Proceedings*, 1879, pp. 51–57; Albany *Cultivator & Country Gentleman*, February 13, 1879.
36 Indiana State Grange, *Proceedings*, 1877, p. 18; 1878, pp. 4–8, 46–47.

College asked the state's lawmakers to issue what amounted to a blank check to the school to enable it to hire more faculty members and to make its agricultural experiment station "first-class." In 1883 the Wisconsin State Grange asked the board of regents of the University of Wisconsin to earmark some money for a state weather-reporting service to be run by the college of agriculture for benefit of farmers.[37]

Although state granges usually did not ask for specific appropriations, there were exceptions. In some way, the Mississippi State Grange calculated in 1880 that it would cost $125,000 to place Mississippi A. & M. College on a firm footing. The order's budgetary request was made in 1880 before the first student entered a classroom at the newly chartered land-grant college. By what means grangers knew in advance how much money it would require to open the doors of the school is a mystery which cannot be answered with certainty, but there is one very feasible explanation. Possibly the board of trustees or President Lee collaborated with the state grange in estimating fiscal needs of the college. Eight of the nine original board members and Lee belonged to the farmers' organization at the time.[38]

In a similar case involving the Wisconsin State Grange, Professor William Henry, a patron himself, more than likely informed one of his brothers that he desired a $12,500 grant from the legislature for carrying on research work at the experiment station farm in Dane County. If he did not, a question immediately arises: How else would the farmers have known his exact needs? In addition, Henry's record demonstrates that he had a knack of coyly dropping suggestions without provoking the University of Wisconsin and its friends in the general assembly.[39]

There was one educational demand made by the Grange which was very consistent with the order's bylaws and constitution. The organization defied tradition throughout the late 1880s by asking for the elimination of all sex barriers at land-grant colleges. Michigan Master C. G. Luce captured the spirit of the organization's feelings on the controversial issue when he raised objections to the way women were being systematically barred from the state industrial college at East

---

37 Texas State Grange, *Proceedings*, 1882, pp. 6, 54–56; Illinois State Grange, *Proceedings*, 1898, pp. 41–42; Wisconsin State Grange, *Proceedings*, 1883, p. 59.
38 Mississippi State Grange, *The State Grange and A. & M. College*, 2–3; Bettersworth, *People's College*, 28, 41–42; Charles E. Rosenberg, "Science, Technology, and Economic Growth: The Case of the Agriculture Experiment Station Scientist, 1875–1914," *Agricultural History*, XLV (January, 1971), 7–10.
39 Wisconsin State Grange, *Proceedings*, 1882, p. 60.

Lansing: "They [ladies] are practically excluded from the College for want of suitable accommodations. Our sons and daughters are reared and educated together in the family, the common school . . . and can anyone give a valid reason why provision should not be made for farmers' daughters at the College as well as for their sons?" [40]

Although stubborn conservative elements in society cringed at the thought of having state-supported coeducational colleges in their midst, the patrons' drive still produced some noticeable results in several states. Of course, the success of the movement was not due entirely to grange efforts; other reform-minded citizens also felt the woman's place was not necessarily limited to the home. With their help, the order was able to effect changes in at least three states.[41]

Of these, the long battle in Michigan was the most dramatic. The state grange and others interested in reform struggled there for more than twenty years before lawmakers finally succumbed in 1897 and passed a law enabling the state's land-grant college to build a girls' dormitory and permitting it to offer courses in "household economy." [42]

Meanwhile, concessions were easier to secure in Connecticut and in Kelley's home state. Patrons did not have to push as long in these states to get their assemblies to approve their requests. The grange crusade in Minnesota lasted eight years. Finally, in 1897 Minnesota lawmakers enacted a law calling for construction of a $25,000 women's residence hall. Interestingly enough, the New Englanders had asked their state assembly to order Storrs to admit girls five years earlier, and after a lapse of only three years one out of six pupils attending the institution was a woman.[43]

The "household economy" courses sought by grangers for their daughters were not the only additions to college curriculums which the order wanted. In 1899 the National Grange also endorsed Presi-

[40] Texas State Grange, *Proceedings*, 1884, p. 18; Boss, *The Early History of the School of Agriculture at St. Paul*, 63–69; Connecticut State Grange, *Proceedings*, 1892, p. 75; 1895, p. 60; National Grange, *Proceedings*, 1874, p. 58; Eleanor Flexner, *Century of Struggle: The Woman's Rights Movement in the United States* (Cambridge, 1959), 122; W. J. Beal, *History of the Michigan Agricultural College and Biographical Sketches of Trustees and Professors* (East Lansing, 1915), 228–29.

[41] Flexner in *Century of Struggle* highlights the handicaps faced by those trying to lift the roadblocks in the way of female emancipation.

[42] Michigan State Grange, *Proceedings*, 1877, p. 25; Madison Kuhn, *Michigan State: The First Hundred Years, 1855–1955* (East Lansing, 1955), 124; Michigan State Board of Agriculture, *Annual Report*, 1897, p. 485.

[43] Boss, *The Early History of the School of Agriculture at St. Paul*, 63–69; Connecticut State Grange, *Proceedings*, 1892, p. 75; 1895, p. 75.

dent Charles William Eliot's improvisions at Harvard and asked that all land-grant colleges adopt his elective system. In order to make their case for his innovation more inviting, patrons prefaced their plea with the old familiar "Educational Allegory" about handicaps confronting different members of the animal kingdom when each was expected to perform the same tasks with no allowance for individual talents inherent in each beast. To them, a direct parallel existed between expecting fish to fly like birds and asking all students to excel at the same subjects. It was because of this conviction that the order felt that colleges should take different aptitudes of their charges into consideration by providing them a choice of courses suited to every ability. According to these farmers, inserting such dispensations into college programs would make higher education a more meaningful experience. This demand followed closely one offered at a preceding session of the National Grange. At that time, grangers asked that foreign languages no longer be required for college degrees. If, on the other hand, students expressed a desire to take one, they should be encouraged to learn Spanish, a practical tongue, and not Greek, a dead and useless language.[44]

These demands were tied closely to others which had been made earlier, calling for a loosening of college entrance requirements. Members of the Grange felt that agricultural colleges should "admit students to the freshman class from the country public schools" because many rural youngsters seeking admission had neither means nor opportunities to attend city high schools.[45]

Grangers also believed that part of every college of agriculture education should be devoted to work experience because students could "secure that physical development now so often sought after in the gymnasium, baseball, and boating clubs" from the learning experience gained by laboring on the college farm. Master Leonard Rhone of Pennsylvania and others in the order concluded that agricultural theory was not enough to prepare young men for farm life. Rhone even went so far as to add that one could easily gain as much from an "apprenticeship to some industrious and prosperous farmer" as he could from an education stressing theory at the expense of practice. Ezra Carr, a former agriculture professor at the University of California and a past master of the Tennessee State Grange, stated the

44 National Grange, *Proceedings*, 1899, pp. 161–62; 1898, pp. 123–27.
45 *Ibid.*, 1898, p. 123; Michigan State Board of Agriculture, *Annual Report*, 1897, p. 489.

case for practical education as well as anyone in the organization when he asked whether it is "so much greater an accomplishment to say horse in half a dozen languages, than to know how to breed and care for one, until the beast has become more than half human in his beauty and intelligence?" Some even carried his line of reasoning one step further by suggesting that all students, "whether born in poverty or affluence, should be taught how to work."[46] Thus, the puritan ethic of hard work and simple living was being revived in grange halls, even if it had little or no impact on developers of college education philosophies.

Ideally, colleges of agriculture in the late nineteenth century should offer two distinct courses of study: a short course for part-time students and a long course for full-time pupils. At least, this was what the Grange thought. By having both, land-grant institutions could cater to needs and time schedules of practically all rural residents seeking to enrich their knowledge of scientific farming techniques. The concept of having two programs of study rested on the belief that a little knowledge was better than none. Moreover, grangers strongly believed that short courses were a perfect supplement to the training of rural schoolteachers. Short courses gave rural teachers enough background in agriculture to teach young children those elementary principles of botany, chemistry, and entomology that applied to the work of the farm.[47]

In 1899 the National Grange summarized its demands for higher education in one policy statement called an "outline of an ideal agricultural college." According to this synopsis, a college year should be divided into four terms, with "all theoretical and class-room work" being completed between October and March. The remaining six months then should be devoted to practical operations, "either on the college farm under competent instructors, or at the student's home." The college should be operated as a "business enterprise" with every student remunerated for his labor "at the usual price paid for the same class of labor on neighboring farms." Under such an arrangement, students would be able to support themselves without resorting to outside assistance, and "at the same time receive the incalculable benefit from 'doing things' with their own hands." In addition, a col-

---

[46] Horton, *The Grange in Vermont*, 31; Pennsylvania State Grange, *Proceedings*, 1886, p. 12; Ezra S. Carr, *The Patrons of Husbandry of the Pacific Coast* (San Francisco, 1875), 381–83; National Grange, *Proceedings*, 1888, p. 133.

[47] Connecticut State Grange, *Proceedings*, 1896, p. 65; West Virginia State Grange, *Proceedings*, 1891, p. 16.

lege of agriculture should be located on a "good farm, and not be made, in any sense, a side show to a classical college or university." Finally, during spring and summer months instructors' time should be utilized in the farm's operation and in experiment station work.[48]

The "outline" failed to include one duty of an "ideal agricultural college." On other occasions, grangers stated that they expected these schools to provide needed training for prospective teachers of agriculture in common schools. They concluded that such provisions should be made by departments of agriculture to accommodate students attending normal schools who expressed a desire to teach in rural districts. Consequently, there were repeated calls for cooperative programs which would enable teacher trainees to study both at colleges of education and of agriculture. Eventually, such courses were developed, but they did not come into existence on a large scale until after Congress passed the Smith-Hughes Act in 1917.[49]

Another service rendered by the farmers' society was the leadership it gave fledgling land-grant institutions. Countless patrons volunteered service as trustees, while others taught in agricultural colleges. Some members of the organization even became college presidents. In fact, the Grange encouraged its members to interest themselves in these schools and to assist them whenever possible.[50]

There was an underlying factor which explains in part why leaders of the order displayed so much interest in these schools. An unsubstantiated fear of what might happen told them to encourage members to serve on governing boards of land-grant institutions. They were certain that bands of eager "professional and political favorites" were waiting patiently on the sidelines for the opportunity to manipulate these institutions. Farmers had been victimized so many times in the past that they envisioned a conspiracy lurking behind everything associated with them. In final analysis, therefore, it was this

48 National Grange, *Proceedings*, 1899, p. 163.
49 Starkville *Southern Live-Stock Journal*, January 13, 1884; Kansas State Grange, "Report of the Educational Committee," in Kansas State Grange Clippings, vol. I, 1877, taken from the *Kansas Farmer*, Kansas Historical Society, Topeka.
50 C. W. Charlton to Saunders, August 14, 1874, in Saunders Collection; Mississippi State Grange, *The State Grange and A. & M. College*, 1–4; West Virginia State Grange, *Proceedings*, 1893, pp. 6–7; Wayland F. Dunaway, *History of Pennsylvania State College* (State College, Pa., 1946), 87, 101; William F. Hill, *A Brief History of the Grange Movement in Pennsylvania* (Chambersburg, Pa., 1923), 39; Boss, *The Early History of the School of Agriculture at St. Paul*, 22, 29–30; Trump, *The Grange in Michigan*, 24.

paranoiac feeling of a world aligned against them that accounts for their exhortations.[51]

One positive accomplishment of the Grange was its encouragement given work of colleges of agriculture. Those institutions which discharged their Morrill Law obligations in accordance with guidelines laid down by Patrons of Husbandry regularly received recognition and praise from grange orators. While it is impossible to estimate the impact that these kind words of commendation had, there is little doubt that they steered to college some farm boys who otherwise might not have gone.[52]

Moreover, the Grange undertook various specific programs to encourage farmers to send their sons to college. In 1892 the Illinois State Grange passed a resolution setting up a competitive scholarship program. According to provisions of this arrangement, the ten young men making the highest scores on an examination were to receive grants to attend the University of Illinois college of agriculture. Five years later, the Michigan State Grange asked its agricultural college to furnish each rural school district with pictures of its campus and facilities. Members hoped that these photographs would advertise the school and encourage youngsters to go there.[53]

Colleges of agriculture responded favorably to grange efforts in their behalf. Their signs of gratitude took many forms. In Pennsylvania and Michigan, Grange Memorial Dormitory and Mary Mayo Hall were testimonies to the work of the organization. Mayo Hall on the campus of Michigan Agricultural College was named for a sister in the order who distinguished herself by working tirelessly for admission of girl students to the college.[54]

[51] Texas State Grange, *Proceedings*, 1884, p. 56; California State Grange, *Proceedings*, 1890, p. 13; Wisconsin State Grange, *Proceedings*, 1882, pp. 62–63.

[52] New Hampshire State Grange, *Proceedings*, 1880, p. 12; Wisconsin State Grange, *Proceedings*, 1882, pp. 13–15; Michigan State Board of Agriculture, *Annual Report*, 1898, p. 671; 1899, p. 665; Delaware State Grange, *Proceedings*, 1899, p. 17; Connecticut State Grange, *Proceedings*, 1890, p. 9; Illinois State Grange, *Proceedings*, 1875. Normally, the Grange reserved its praise for public institutions, but there was at least one exception. Keuka College of New York was enthusiastically endorsed in 1891 by the state grange. The organization singled out that private liberal arts college because it opened its doors in 1890 to students of both sexes and of all sects. New York State Grange, *Proceedings*, 1891, p. 107.

[53] Illinois State Grange, *Proceedings*, 1892, p. 58; Michigan State Board of Agriculture, *Annual Report*, 1897, pp. 489–90.

[54] Frederick O. Brenckman, *History of the Pennsylvania State Grange* (Harrisburg, 1949), 275–80; Kuhn, *Michigan State*, 354–55.

In 1892 the Grange, in cooperation with Mississippi Agricultural and Mechanical College, erected an obeliscal stone monument on the campus in memory of Putnam Darden, a state and National Grange leader who fought for establishment of the college and afterward defended and encouraged its programs. The pillar bears these words:

Capt. Put Darden, Born Mar. 10, 1836, died July 17, 1888, age—52 yrs. 4 m's., 7 d's. A true Patron. A devout Christian. Darden

Master of State Grange of Mississippi and Master of the National Grange.

"Whether on the scaffold high,
Or in the battle's front
The noblest place for a man to die
Is when he dies for man."

Erected as Loving Tribute by the Patrons of Husbandry of the United States.

In addition, a street on the campus was named after Master Darden.[55]

Occasionally, leaders of the farmers' society were invited to give addresses at these schools. Master N. J. Bachelder of the New Hampshire State Grange was a commencement speaker at the 1895 convocation exercises of Connecticut State Agricultural College, and he was only one of many so honored to speak at institutions to graduates.[56]

College leaders also showed their support for granger activities in the field of higher education by agreeing to speak at grange functions. These addresses usually contained praise for the organization and explanations of their colleges' activities and programs. On other occasions, school officials demonstrated their gratitude by permitting charterings of subordinate granges on their campuses. At Mississippi Agricultural and Mechanical College, two faculty members served as sponsors for the school's chapter. Similar bodies existed at the University of Alabama and West Virginia University.[57]

[55] Mississippi Agricultural and Mechanical College, *Biennial Report of the Trustees*, 1892–93, p. 11; Bettersworth, *People's College*, 113; words taken from the monument by the author.

[56] Concord *New Hampshire Agriculturist and Patron's Journal*, June, 1895.

[57] Topeka, Kansas, *Capital*, December 15, 1881; Wisconsin State Grange, *Proceedings*, 1881, p. 51; Massachusetts State Grange, *Proceedings*, 1898, pp. 69–77; Starkville (Miss.) *Southern Live-Stock Journal*, August 4, 1881; Houston Cole, "Populism in Tuscaloosa" (M.A. thesis, University of Alabama, 1927), 41; Barns, "The Influence of the West Virginia Grange upon Public Agricultural Education of College Grade," 147–48.

Between 1870 and 1900, no other agrarian organization worked as energetically as did the Grange in promoting agricultural education at the nation's land-grant colleges. In their zeal some members in fact went to extremes. Only a slim minority of grangers were really positioned to judge and evaluate complexities of a college program, but nevertheless most patrons were not hesitant to criticize college officials and to demand immediate remedial action whenever they thought it was necessary to safeguard principles of the Morrill Act as they interpreted them.

Especially, did patrons oppose the teaching of anything except agriculture at land-grant institutions, and like most farmers they were more than suspicious of "classical departments." In this respect, grangers were as narrow-minded as they accused their opponents of being. In retrospect, neither side was willing to recognize that the other had a valuable contribution to make to education.

In balance, there can be no doubt that the Grange played a powerful, positive role in the development of land-grant institutions. Its members did much to elevate agricultural training to respectability and to place that training on a solid foundation. In the course of three decades, grangers suggested many beneficial programs and at crucial moments offered both leadership and encouragement to struggling agricultural colleges. For these efforts alone, the Grange earned its place as the major farm organization of the age.

CHAPTER V

The Grange
Adult Education Program

The average farmer of 1867 was not much better intellectually than his ancestor living at the time of the War of 1812. Members of both generations had narrow horizons, suspected strangers, rejected new ideas, and scorned anything scholarly. In fact, most of both generations believed that an elementary knowledge of the three R's was all farmers needed to meet their problems. To these unenlightened souls, any additional learning was a frivolous waste of their time and energy; "education and farming, like oil and water, would not mix." [1]

Founders of the Grange were aware of this widespread attitude, and they also knew that farmers eventually would have to mend their ways or be further isolated from the mainstream of American life. Oliver H. Kelley and his associates grasped what was taking place in the United States. They recognized the urban and industrial trends of their nation, and they realized that rural residents had to be awakened from the lethargy. Consequently, they sought to create an agrarian organization that would broaden the member's perspective, dispel his fears of the unknown, and help him to take his rightful place in the new America that was emerging. [2]

Grange leaders knew they faced a difficult battle when they tackled the problem of farmers' attitudes, but they were confident that their efforts would net favorable results. They developed no strategies for achieving this goal, but a general pattern of action still existed on state and national levels of the organization. Use of slogans, such as "education like the rays of the sun, should descend upon all" and exhortations emphasizing virtues and importance of adult education became regular features of almost every state and national grange session. In other words, grange leaders were almost obsessed with the

[1] Arthur E. Paine, *The Granger Movement in Illinois* (Urbana, Ill., 1904), 46.
[2] Kelley, *Origin and Progress of the Order*, 12–20; Jonathan Periam, *The Groundswell: A History of the Aims and Progress of the Farmer's Movement* (Chicago, 1874), 125.

idea that a "starved mind" was as damaging for an individual as a crippled and diseased body.[3]

When Kelley and his friends formulated a plan for the Grange, they made sure that local meetings would include features tending to link all members with the rest of the nation, and at the same time to improve them through education. With these thoughts in mind, the founders devised a comprehensive format for the grange meeting. It contained a rather complex ceremony suggesting brotherhood and fidelity, but more importantly the local meeting had a special period set aside for education. This part of each session became known as the lecture hour. It was during this time period that members gathered to improve themselves educationally by discussing items of mutual concern.[4]

Constitutions and bylaws of state and national granges, however, set no specific meeting schedules for local chapters. Consequently, some subordinate granges met weekly, whereas others convened twice a month. A few locals chose to have only twelve sessions a year. Meetings were usually held on Fridays and Saturdays. Pomona or county granges held quarterly sessions, generally on Thursdays.[5]

There exists a popular misconception regarding the lecture hour. Since connotations associated with the First Granger Movement have largely been political and antibusiness, most professional historians have assumed that the lecture hour was devoted mostly to discussions concerning the regulation of railroads and to conversations dealing with monopolies and politics. Although these topics certainly were brought up from time to time, they were not primary concerns of grangers. Far more interest was shown in a wide range of matters dealing directly with better farming techniques. In other words, patrons gave priority to immediate and practical agricultural problems and

3 Albany *Cultivator and Country Gentleman*, February 7, 1878. The regularity with which grange leaders addressed themselves to the importance of adult education makes it impossible to list all the sources used for making these observations, but typical examples of their oratory and committee reports can be found in the following: New Hampshire State Grange, *Proceedings*, 1892, pp. 81–85; Washington State Grange, *Proceedings*, 1891, p. 11; National Grange, *Proceedings*, 1878, pp. 99–103; 1879, p. 25; 1889, p. 155; 1894, p. 188.

4 Kelley, *Origins and Progress of the Order*, 12–20; Jennie Buell, *The Grange Master and the Grange Lecturer* (New York, 1921), 18–20; National Grange, *Manual of Subordinate Granges* (Philadelphia, 1874).

5 Rhode Island State Grange, *Proceedings*, 1894, p. 15; New York State Grange, *Proceedings*, 1879, pp. 108–17; Wisconsin State Grange, *Proceedings*, 1890, pp. 56–57.

not to matters affecting them only commercially and politically.[6]

Available evidence shows that grangers concerned themselves primarily with agricultural and household topics. Male members studied such things as fertilizers, cotton, tobacco cultivation, meat preservation, potato bugs, fruit culture, and a wide range of other items dealing with the maintenance of a productive farm. When ladies conducted the lecture hour, they discussed cooking, baking, canning, gardening, sewing, and embroidering.[7]

Although it was the lecturer's duty to plan the educational portion of each meeting, he usually did not have to rely exclusively on his own imagination. State and national granges generally provided suggestions for possible discussion topics. These hints were listed in many grange periodicals, circulars, and proceedings. The Ohio State Grange printed a lengthy manual entitled *Hints and Helps*, which suggested themes for discussions and questions appropriate for debates. Under the "agricultural" heading, members were told what points to include in talks on a variety of subjects, including drainage, corn, wheat, potatoes, fertilizers, fencing, timber, and grasses. For inquiries concerning the latter, members were urged to comment on such aspects as "Varieties: for permanent pasture; for rotation; for hay.—Time and method of seeding.—Time for cutting hay.—Under what circumstances is it more profitable to sell hay?—Under what circumstances more profitable to feed?" The debate section told patrons how to prepare their side of controversies, listed topics that lent themselves to two-sided discussions, and gave hints for increasing audience participation. The last part of the guide listed a model lecture program for a full year's schedule.[8]

[6] Solon Buck perpetuated this myth by tying the organization so closely to commerce, independent political parties, and to fights against monopolies. See Buck, *The Granger Movement*.

[7] Bloomington Grange (Minn.) No. 482, Calendar, 1896, Oliver H. Kelley House, Elk River, in Minnesota, Papers, Minnesota State Historical Society, St. Paul; North Star Grange, Secretary's Minutes, September 12, 1868; Wilmington Grange (Fluvanna County, Va.), Secretary's Minutes, August 1, 1875, October 6, 1877, in University of Virginia Library, Charlottesville; Hiawatha Grange (Hiawatha, Kan.) Secretary's Minutes, June 22, 1872, Kansas State Historical Society, Topeka; Borrors Corners Grange (Jackson Township, Ohio) No. 608, Secretary's Minutes, June 3, 1876, Ohio State Historical Society, Columbus; Clarke County (Ga.) Grange No. 101, Secretary's Minutes, October 7, 1873, February ?, 1876, Clarke County Grange Records, University of Georgia Library, Athens; Olive Grange, Secretary's Minutes, January 9, 22, 1887, February 21, 1891; Church Hill Grange, Secretary's Minutes, January 26, February 9, 23, March 9, April 24, May 21, 1874.

[8] National Grange, *Proceedings*, 1885, p. 131; Springfield, Ohio, *Grange Visitor and Farmers' Monthly Magazine*, January and February, 1875; Ohio State Grange,

Many locals placed "query boxes" in their lodge halls to give members a voice in their meetings. The topics which arose from this source often were quite beneficial. The query box in Indiana's Olive Grange No. 189 yielded two typical questions February 25, 1888. One dealt with whether the government had constitutional authorization to sell land, and the other questioned whether sugar prices would drop if sugar taxes were abolished.[9]

The grangers also used other educational techniques. Committees and individuals were appointed to assess members' farms. They checked for evidence of soil erosion and depletion, noted the types of crops grown, inspected farm buildings and fences, and viewed livestock. They presented a report of their findings at the next meeting and gave analyses of each farm visited and offered suggestions.[10]

Kentucky's Church Hill Grange chapter was more concerned with how members were utilizing their land, so it delegated a brother to visit each farm and record how each acre was used. His investigation showed among other things that these patrons were practicing crop diversification:

Table 13

CROP DIVERSIFICATION
AMONG CHURCH HILL GRANGERS

Farm Number	Acres in Corn	In Wheat	In Tobacco
1	45	45	11
2	21	90	14
3	25	50	9
4	100	45	30
5	130	190	80
6	140	250	75
7	31	40	21
8	40	80	14
9	18	55	10
10	50	80	24

Confidential Trade Circular, 1895, pp. 56–63; Denver *Colorado Farmer*, January 24, 1884; Ohio State Grange, *Hints and Helps to Profit and Pleasure in the Grange with Topics for Discussion and Programmes for Meetings* (Springfield, Ohio, 1881), 17–33.

9 Douglas County Pomona Grange, Secretary's Minutes, March 25, 1876; Olive Grange, Secretary's Minutes, February 25, 1888.

10 Chicago *Prairie Farmer*, February 26, 1876.

Table 13— (Cont.)

CROP DIVERSIFICATION
AMONG CHURCH HILL GRANGERS

Farm Number	Acres in Corn	In Wheat	In Tobacco
11	50	70	19
12	80	80	23
13	30	55	25
14	40	60	50
15	25	25	13
16	45	52	16
17	30	38	12
18	23	18	55
19	80	70	45
20	25	90	20
21	25	50	16
22	9	30	9
23	15	40	10½
24	45	45	17
25	75	60	23
26	14	27	7
27	20	60	38
28	45	40	37½
29	40	60	20
30	30	13	11
31	40	60	22

These statistics also highlight the fact that the lecture hour was a very democratic institution in that it brought together farmers from all economic backgrounds.[11]

Many topics covered in grange halls had only regional importance. For example, order records lucidly show that many southern farmers understood their economic problems. Granges often traced their members' woes to reliance upon staple crops. Patrons realized that the world production of certain commodities was so much in excess of consumption that prices could only decline as long as farmers continued to grow these crops.[12]

[11] Church Hill Grange, Secretary's Minutes, October 12, 1877.

[12] Louisiana State Grange, *Proceedings*, 1875, p. 13; Booneville (Ark.) *Enterprise*, August 6, September 3, 1875; South Carolina State Grange, *Proceedings*, 1875; Clarke County Grange, Secretary's Minutes, October 1, 1873; Atlanta *Wilson's Herald of Health, and Farm and Household Help*, December, 1873.

Besides isolating the source of their afflictions, grangers knew how to remedy problems of overproduction. In the Cotton Belt, state and local chapters repeatedly urged their members to diversify and to plant crops other than cotton. Throughout Dixie, in the fall and winter patrons pledged to balance their fields during the next growing season with cotton, grains, and grasses. At the 1875 session of the Georgia State Grange, delegates overwhelmingly adopted a resolution introduced by Leesburg Grange No. 16, requesting the state body "to use all of its moral influence in impressing upon the planters and farmers of our state the absolute necessity of so curtailing the prospective cotton crop and amplifying the supply of provisions, grasses and fruits, that we may once more be relieved of the fearful incubus of poverty and distress that paralyzes every energy and overshadows the future with clouds of doubt and darkness." Two years earlier on December 15, 1873, members of the state's Clarke County Grange No. 101 met the problem squarely by resolving to "plant one third of our crop in 1874 in small Grain=one third in corn and the remainder in Cotton." [13]

At its 1874 conclave the National Grange turned attention to diversification. Nine masters from the South explained to the rest of the delegates how excessive production of cotton was depressing the price and recounted how farmers and planters added to their miseries by growing more instead of less cotton. These amateur economists predicted that prosperity would return to Dixie as soon as the supply was cut to a level below the needs of the world's textile mills. This could be done, they maintained, by empowering the National Grange to set a quota of 3,500,000 bales per year and giving it enforcement authority to insure that no more was grown.[14]

In 1874 State Agent Henry Lee of Colorado made speeches throughout the territory, urging farmers to diversify. He deplored the fact that the region was losing thousands of dollars because of its farmers' refusal to grow corn, oats, barley, and forage crops and their reluctance to keep poultry, swine, and cattle on their farms. He noted that in 1873 Colorado had imported $375,000 worth of grains and feeds and $1,000,000 worth of meat, eggs, and dairy products. In his speeches, Lee lampooned farmers for causing this drain on the territory's economy and pleaded with his audiences to reduce or eliminate

13 Georgia State Grange, *Proceedings*, 1875, p. 18; Clarke County Grange, Secretary's Minutes, December 15, 1873.
14 National Grange, *Proceedings*, 1874, pp. 60–63.

this outward flow by planting a wide range of crops and by having barns, pens, and coops filled with cattle, pigs, and chickens.[15]

To complement its pleas for diversification, the Grange in some areas established incentive programs. Awards were given for outstanding grain and vegetable samples and for healthy livestock specimens. A grange near Kington, North Carolina, had almost $200 in its premium chest, and it gave cash prizes for noteworthy agricultural products. In the heart of the Cotton Belt, a Mississippi chapter offered $25 to the member producing the largest yield of corn per acre and $15 to the patron producing the most syrup from one acre of cane. It also granted $10 for the fattest hog. Ironically, some granges offset the influence of these diversification projects by giving bonuses for cotton. A local chapter at Banks, North Carolina, did so in 1874, offering $25 to the producer who grew the most bales per acre.[16]

Discussing problems connected with staple production helped farmers to understand what needed to be done, but it had little or no effect on national output of these crops. Statistics for the period when the Grange thrived in the South show the ineffectiveness of the order's efforts to reduce overproduction by consent. Between 1872 and 1882, the number of acres planted in cotton soared from 8,500,000 to 16,800,000, and the number of bales produced multiplied from 3,900,000 to 6,900,000.

When extra money was available in grange treasuries, it was quite common for masters and lecturers and occasionally other officers of state and national granges to spend a considerable part of their time visiting subordinate chapters, encouraging them and spreading information generally.[17]

The best meetings were not necessarily those addressed by professional grange orators. In fact, some of the most beneficial and "useful" sessions resulted from lecturers' ingenuous improvisations. In at least two instances, new life was injected into local granges when members brought products of their farms and work benches with them to meetings. A sense of accomplishment resulted from such meetings; sisters and brothers who belonged to the order were always proud to display

---

[15] Denver *Rocky Mountain Weekly News*, March 4, 1874.

[16] Raleigh (N.C.) *State Agriculture Journal*, May 27, 1875; Charleston *Rural Carolinian*, June, 1874.

[17] U.S. Department of Agriculture, *Cotton and Cottonseed: Acreage, Yield, Production, Disposition, Price, and Value, By States, 1866–1952*, Agricultural Marketing Service, *Statistical Bulletin* 164 (Washington, 1955).

their products. Patrons also relished spirited competition; they liked any meeting that included rivalry. When the lecturer of Wisconsin's Prairie Grange No. 203 arranged a debate by dividing the chapter into teams, all members enjoyed presenting a part of the program. After each participant had been given an opportunity to expound, judges evaluated performances of each squad and then ruled that the losers had to serve refreshments to the winners. Since the intent of the program was joviality and insight, no one left the meeting with bitterness.[18]

Responses from a New York State Grange questionnaire sent to every local chapter in the state showed that members felt that they benefited more from the lecture hour than from any other aspect of the order's program. In this probe, subordinate grange masters had been asked to answer twelve questions about the organization and its activities. Of these, four had been directly related to the lecture hour. To these questions, local officials overwhelmingly responded that the time set aside for education was producing satisfactory results. In reply to the inquiry, "Have the farmers and their families who belong to your Grange been led to read and think more because of their connection with the order?," almost every master stated "they have." [19]

Institute speakers frequently said that they could "always tell . . . a community where a Grange exists" because they could spot the patrons in any audience. Grangers were conspicuous because they were more relaxed in front of audiences, were better read, and were more knowledgeable of parliamentary procedure than those outside the order. Public servants frequently claimed that social and educational changes taking place were the result of the lecture hour. Even if politicians had selfish motives for making public their evaluations, their summarizations contained much truth about the effectiveness of the grange program. For example, no one could challenge Lieutenant-Governor Alexander Pancoast Riddle of Kansas who, at the 1885 session of the state grange, stated that meetings of fraternal orders served as "valuable schools for those who attend[ed] them." Riddle conjoined this statement with one crediting the order with turning "the most diffident men" into "the most able and logical debaters." He concluded his remarks by saying that the evidence showed that the

18 Charleston *Rural Carolinian*, January, 1875; New York *American Agriculturist*, September, 1877; Wisconsin State Grange, *Proceedings*, 1891, p. 23.
19 New York State Grange, *Proceedings*, 1877, pp. 73–77.

grange meeting "awakens investigation and promotes a desire for learning." [20]

A minister observed that "since the introduction of the Grange, I have seen a remarkable change in the walk and conversation of my flock; they are more careful in their dress and general appearance and are reading more." A postal clerk in South Carolina observed that "there are now thirty newspapers taken at this office, whilst there was but one taken before the establishment of the Grange in this vicinity." [21]

Agricultural journals were indeed flooding rural post offices, furnishing perhaps the best evidence supporting a grange claim that the lecture hour was affecting farmers' lives. Scores of farmers who had neither seen nor read these magazines before the advent of the Grange now subscribed to them. In fact, there were so many new readers that the order caused a mild literary revolution. At the height of the First Granger Movement, numerous new agricultural journals appeared, doubling the number of such periodicals in the United States, and newspaper and magazine subscriptions multiplied. Many of the new magazines were established with the expressed purpose of capturing granger attention.[22]

And as one might expect, opportunists soon recognized the potential that the new market presented. When the order first appeared, editors and publishers often reserved judgment and hesitated to endorse the new farmers' fraternity. However, as soon as it became evident that the Grange would succeed and attract members by the thousands, agricultural journalists and rural newspaper editors rushed to express positive sentiments of the Patrons of Husbandry. Most editors who depended upon rural trade began catering to grangers by printing more and more news about them and their activities. In some cases, there were complete editorial reversals. Skeptics suddenly became supporters, and critics turned into defenders. When owners William and W. L. Jones of the *Southern Cultivator* first learned of the organization, they sensed a northern conspiracy and they warned

20 Kenyon L. Butterfield, "The Grange," *The Forum*, XXXI (April, 1901), 236–37; J. Harold Smith, "History of the Grange in Kansas, 1883–1897" (M.A. thesis, University of Kansas, 1940), 182–183; Denver *Colorado Farmer*, March 13, 1884; Kansas State Grange, *Proceedings*, 1885, p. 27.

21 D. Wyatt Aiken, "The Grange: Its Origin, Progress, and Educational Purposes," United States Department of Agriculture, *Special Report* 55 (Washington, 1883), 10.

22 Homer Clevenger, "Agrarian Politics in Missouri, 1860–1896" (Ph.D. dissertation, University of Missouri, 1940), 77–79.

their readers of the impending threat that the order posed to the South. As late as February 1, 1872, they asked if southerners had had " 'entangling alliances' enough already with the North, to warn us against going blindfold into an organization with a great *central head* at Washington City, exacting tribute from local organizations all over the country." This fear, however, disappeared completely after the Grange had enrolled Dixie farmers by the thousands. The Joneses obviously realized that they had to recant if they were going to compete with progrange journals.[23]

In many states, the Grange adopted one publication as its official organ. These journals were usually either owned by the order or were edited by prominent members, and they generally had either Grange or Patron in their titles. Of course, when a state grange chose one paper over others as its official publication, bitterness flared between overlooked journals and leaders of the organization. The staffs of the *Kansas Farmer* and the *Farmers Home Journal* of Kentucky were especially disgruntled after their publications had been by-passed. In fact, editors of the latter were so angry that they urged their readers who belonged to the Grange not to pay their dues: "If the Patrons pay dues to the ring that controls the State Grange in this State about $700 will go to publishing a miserable apology for a something which will . . . be called a Bulletin." [24]

Being selected as an official organ offered publishers certain advantages, but the effect did not usually last very long. Grangers' fickleness brought most of these publications to disaster after only a few months' printing. The problems that plagued the *Spirit of Kansas* were typical of those experienced by other organs. Editor James S.

23 Baltimore *American Farmer and Rural Register*, October 1, 1873, October 1, 1874; Charleston *Rural Carolinian*, March–September, 1871; Indianapolis *Indiana Farmer*, 1874; Clarkesville (Tenn.) *Tobacco Leaf*, July 29, 1874; Mobile *Rural Alabamian*, July, 1873; Des Moines *Iowa Homestead and Western Farm Journal*, January 20, 1871; Madison *Western Farmer*, April 12, 1873; Fayetteville (Ark.) *Democrat*, December 20, 1873; Athens (Ga.) *Southern Cultivator*, February 1, 1872, May, July, 1873.

24 Michigan State Grange, *Proceedings*, 1875, pp. 37–38; Minneapolis *Farmers' Union*, April, 1871; National Grange, *Proceedings*, 1882, pp. 105–106; Macon *Missouri Granger*, March 10, 1874; Maine State Grange, *Proceedings*, 1875, p. 41; Cleveland *Ohio Farmer*, July 22, 1897; Charleston *West Virginia Farm Reporter*, December, 1897; North Carolina State Grange, *Proceedings*, 1878, pp. 16, 23–24; Richmond *Virginia Patron*, May 4, 1877; Little Rock *Arkansas Grange*, May, 1874; Indianapolis *Indiana Farmer*, March 11, 1876; Lawrence *Spirit of Kansas*, August 9, 1873; Leavenworth *Kansas Farmer*, January 7, 1874; Madison *Western Farmer*, January 24, 1874; Louisville *Farmers Home Journal*, February 21, 1878.

Stevens complained that his paper was "fearfully hard up" long before the journal celebrated its second anniversary. Maintaining that not enough patrons had subscribed to the paper to make it financially solvent, Stevens moaned that no one but grangers themselves were at fault for the woeful plight of the *Spirit*.[25]

In their attempts to corner the lucrative grange market, publishers often found it necessary to make special offers to granges. Subscription rate wars resulted in some areas. Local chapters which kept abreast of what was happening frequently could take advantage of these windfalls by soliciting bids before making any purchases. This procedure helped them to stock their lodge reading rooms at sizable savings since many companies obviously thought sales were more important than rigid adherence to rules against discounts.[26]

An examination of grange organs reveals why so many farmers refused to renew their subscriptions. These publications not only were repetitious, but they contained very few worthwhile articles. Their pages were cluttered with charts listing local grange officers, songs and poems extolling virtues of the order, long articles justifying almost every action taken by the Grange, serialized short stories, and fashion and household hints. Essays on scientific agriculture were usually reprinted from older journals like the *Rural New Yorker* and were overshadowed by organization topics, such as why the Grange remained a secret society. It seemed that every other issue took up this question.[27]

For the most part, the low quality of these magazines can be attributed to one cause: inexperience. Grange organs were generally managed and edited by novices who blindly entered the field of journalism "to help build up and permanently establish a *purely* Grange paper" in their state. A good case in point involved the Roanoke District Grange of North Carolina. On September 3, 1878, its members decided that the state grange needed a voice, so they created a stock company and purchased the presses of a defunct newspaper. Since no one in the grange had any newspaper experience, the patrons decided that

25 James B. Stevens to Samuel N. Wood, February 4, 1874, in Samuel Newitt Wood Papers, Kansas State Historical Society, Topeka.

26 Easterby, "The Granger Movement in South Carolina," 27; Louisville *Farmers Home Journal*, February 20, 1879; Hiawatha Grange, Secretary's Minutes, June 8, 1872.

27 For examples of Grange journalism, see Little Rock *Arkansas Grange*, May, 1874; Springfield (Ohio) *Grange Visitor and Farmers' Monthly Magazine*, September, 1875.

their secretary and general deputy, R. I. Beale, was the most logical choice for the post of editor and publisher of the proposed journal. Having by his own confession no "experience or scarcely the least idea of conducting a newspaper office—to—say nothing of the printing," Beale was still imbued with enough courage and idealism to accept the challenge and to embark on a new career as general manager of the *Roanoke Patron*. Before he commenced his duties, Beale joked about prospects of the *Patron*, predicting "for it a career as ephemeral as its predecessors." Although made in jest as a witticism, his prophecy nevertheless was fulfilled after only a few issues of the *Patron* had come off the presses in 1879.[28]

By 1880 many grange-owned organs had already died from "want of patronage" and from lack of interest. Less than one-third of the ten thousand grange families in Michigan were subscribing to their state grange's voice, the *Grange Visitor*, in 1877.[29] When the circulation level plummeted below the break-even line, state granges reluctantly had to abandon their publishing enterprises.

Of course, reading materials were essential to the well-being of a successful grange. In order to prepare stimulating programs, members had to have access to suitable books, pamphlets, and periodicals. Since few rural communities had book-lending depositories and since not many farmers collected books as a hobby, many granges set aside space in their temples for reading rooms and began stocking them with a variety of books, magazines, almanacs, and technical bulletins. In most cases, patrons donated whatever books they had, and other volumes were secured with excess treasury funds. A chapter in Marshall, Michigan, raised $500 for its library. Granges acquired transactions and reports from experiment stations, horticultural and agricultural societies, and state and national departments of agriculture. Grange cells also subscribed to various agricultural journals; and in Ohio, granges were able by 1897 to supplement their own materials with books secured from the state circulating library. By the end of the year, at least thirty chapters had taken advantage of this service.[30]

28 Roanoke District Grange, Potecasi (N.C.) Circular from the Secretary's Office, May 15, 1879, R. I. Beale to Polk, May 9, 1879, July ?, 1879, both in Polk Papers.
29 South Carolina State Grange, *Proceedings*, 1877, p. 5; Church Hill Grange, Secretary's Minutes, July 23, August 12, 1875; Michigan State Grange, *Proceedings*, 1877, p. 51.
30 Indianapolis *Indiana Farmer*, June 1, 1878; Hiawatha Grange, Secretary's Minutes, June 22, 1872; North Star Grange, Secretary's Minutes, August 2, 1873; Bloomington Grange Library Records, in Oliver H. Kelley House Papers; West

Many grange libraries were operated under the same principles as public book-lending institutions. Members were fined if they kept volumes longer than the allotted period, usually the time-span between sessions. Moreover, patrons who accidentally damaged or lost books were assessed accordingly. Local chapters generally charged their secretaries with the responsibility of keeping library records and collecting fines. Any revenue received from overdue books or from losses or damages usually went for purchase of additional volumes. Since encouraging farmers to read more was a vital facet of the grange adult education program, national and state granges saw many benefits accruing from these local chapter libraries. Consequently, every encouragement was given. Granges without libraries were urged to establish them. Subordinate chapters also received advice on how to secure books and how to manage reading rooms.[31]

Granges not only encouraged their members to read more, but they occasionally outlined what patrons needed to read. In 1887 the Massachusetts State Grange broadened its adult education program by establishing the "Grange Course of Reading and Study." Under it, members who elected to take "the course" were directed to read certain books, make reports, and have discussions. The program began in 1888, and books selected for that year included: Peabody's *Hand-Book of Conversation;* Blaisdell's *First Steps with American and British Authors; Essays and Correspondence*; Meservey's *Book-Keeping, Single Entry*; Waring's *Elements* (of agriculture); Barnes' *General History*; Laughlin's treatise on political economy; Shaler's *First Book* (on geology); and Mrs. Lincoln's *Boston Cook Book*. In addition students who subscribed for the course had to read and discuss several experiment station bulletins.[32]

Within a few years, news of the Massachusetts Grange Course of Reading and Study spread to other states. In 1889 members of Indiana's Olive Grange No. 189 formed a "reading circle," and two years

---

Butler Grange, Records of the Grange Library; Charleston *Rural Carolinian*, June, 1874; Association of American Agricultural Colleges and Experiment Stations, *Proceedings*, 1889 (Washington, 1889), 89; Delaware State Grange, *Proceedings*, 1899, p. 16; Albany *Cultivator and Country Gentleman*, February 10, 1876; Excelsior Grange (Nebr.) No. 26, Secretary's Minutes, March 23, 1877, in Excelsior Grange Papers, Nebraska State Historical Society, Lincoln; Ohio State Grange, *Proceedings*, 1897, p. 11.

31 West Butler Grange, By-Laws of the Grange Library, April 8, 1876; National Grange, *Proceedings*, 1878, p. 100; 1885, pp. 130–31; New Hampshire State Grange, *Proceedings*, 1891, p. 123.

32 Massachusetts State Grange, *Proceedings*, 1888, p. 81.

later the New York State Grange appointed a special committee to set up a similar program in New York. Members of the New York committee drafted a lesson plan which included sections on American geography, literature, government, and history. In early 1894 the New Hampshire State Grange adopted with some deviations the Massachusetts plan, but instead of having a committee assign what was to be read, they delegated this authority to the state lecturer. He prepared syllabi for each of the topics covered and forwarded them to "each member of the class." In the first year, six hundred patrons representing fifty-seven subordinate chapters completed the course, which included lessons in "Origins of Soils," "Botany," "Plant Growth," "Entomology," "Political Economy," and "Parliamentary Law." New Hampshire leaders expressed some dissatisfaction with the initial response and results, but they were confident that the program would succeed.[33]

Some farmers and their wives complained that reading assignments were too difficult. One New Yorker who had taken his state grange's reading course commented that he was "thick-headed, and it is hard for me to make much progress in anything technical. My 'Plant Life on the Farm' [one of the readings] is beginning to look like my old spelling book that I never mastered. I think your course of reading is on too high a key for the average farmer." Another enrollee who had taken the same course had no trouble understanding the material, but felt the course lacked direction. She thought the program would be strengthened if "a live teacher" supervised it.[34]

The agricultural publications were often too technical for laymen to understand. Consequently, members of the grange hierarchy used every opportunity to ask the scientists writing bulletins and reports to make them less complicated. At the 1889 convention of the Association of American Agricultural Colleges and Experiment Stations, Master S. H. Ellis of the Ohio State Grange explained how writers of experiment station and government bulletins could make their publications sufficiently "attractive" to increase farmers' reading:

> Let them be varied, and let them be small, just a few pages.... The farmer is scared by large volumes. ... Get the bulletins up in attractive form. I remember years ago reading an old book with a peculiar title.

[33] Olive Grange, Secretary's Minutes, December 21, 1889; New York State Grange, *Proceedings*, 1892, pp. 81–86; New Hampshire State Grange, *Proceedings*, 1894, pp. 17–18.

[34] New York State Grange, *Proceedings*, 1894, pp. 126–30.

There was a picture of a gladiator with a drawn dagger, and under it the words, 'Read me, or I will stab you.' Send off your bulletin with something like that. It will attract the attention of the farmer. He will ask, 'What is this that is so blood-thirsty?' . . . The farmer will become interested, will call the attention of his wife to it, and will ask John to read it.[35]

Grange reading courses eventually disappeared because land-grant institutions began providing home extension services. Michigan Agricultural College at East Lansing was a pioneer in the field. With assistance from the state grange, it initiated its first program of correspondence courses in 1891. Thereafter, the Michigan State Grange encouraged its members to enroll in the program and to organize study groups. So successful was the Michigan program that the National Grange resolved in 1898 and 1899 that it "heartily favor[ed] the establishment of correspondence courses of instruction in agriculture, in all . . . agricultural colleges, and . . . [it] urge[d] every member of the Grange and the farmers generally to avail themselves of such instruction." [36]

Active granges supplemented their education programs with farmers' institutes, fairs, and encampments. In the 1870s, order-sponsored institutes were planned and staffed by grangers with no aid from agricultural colleges. These events often lasted two days and featured lectures delivered by local as well as by guest orators and discussions on a variety of topics affecting the farmer and his family, including temperance, middlemen, woman suffrage, and many subjects related to the operation of a successful farm. Granges had three main purposes in mind for planning these sessions. Institutes were supposed to stimulate thought, foster friendships, and improve the farmers' lot by acquainting them with better agricultural techniques.[37]

The quality of grange institutes improved steadily after the organization became affiliated with the land-grant college movement. As members of the order championed colleges of agriculture, administrations of these institutions showed their appreciation by directing professors to carry their lessons of the classroom and laboratory to the farmer. President Stephen Dill Lee of Mississippi Agricultural and

35 Association of American Agricultural Colleges and Experiment Stations, *Proceedings*, 1889, pp. 41–42.

36 Michigan State Grange, *Proceedings*, 1891, p. 53; National Grange, *Proceedings*, 1898, p. 123; 1899, p. 159.

37 Chicago *Prairie Farmer*, January 9, 1875; Chicago *Industrial Age*, May 1, 1875. For the full development of agricultural extension, see: Roy V. Scott, *The Reluctant Farmer: The Rise of Agricultural Extension to 1914* (Urbana, 1970).

Mechanical College was one administrator who responded to the friendly overtures of grangers. He announced in the spring of 1884 that he was going to send a team of professors to grange halls across the state to present lectures and to lead discussions on scientific agriculture. In 1883 Professor W. J. Beal of Michigan Agricultural College toured rural areas of his state and of northern Indiana, addressing grangers.[38]

During the Second Granger Movement, the grange institute program entered a new phase. Land-grant institutions exerted more and more influence in planning and staffing these extension activities. Some of this can be attributed to patrons' demands for greater college participation. In 1890 the California State Grange appealed to regents of the state university to appoint an agricultural expert who would "devote his whole time in the field." At their spring meeting, regents "in compliance with the request of the State Grange for the appointment of a Lecturer to take charge of the holding of 'Farmers' Institutes' in various agricultural centers of this State," chose Professor E. J. Wickson "to organize and carry into effect a system of farmers' meetings." [39]

Where agricultural colleges already cooperated with granges in conducting institutes, the order looked for ways to improve existing extension programs. In Michigan at the 1891 session of the state grange, the committee on education expressed some dissatisfaction with the extent of the institute program. Committeemen were disappointed that every county in the state had not held an institute during the previous year. They pointed out that professors from Michigan Agricultural College had been available and that granges had only to take advantage of this service. Meanwhile, in Kansas, the state grange called upon the Board of Agriculture and the agricultural college to expand their institute activities by adding to the staff already in the field experts in butter and cheese manufacture, sugar production, silk culture, and other assorted agricultural arts.[40]

In general, granges were well satisfied with cooperative institute programs in their states. The Maryland State Grange was pleased with results flowing from the extension work sponsored by Maryland Agricultural College and Experiment Station in conjunction with

38 Starkville *Southern Live-Stock Journal*, May 1, 1884. Indianapolis *Indiana Farmer*, February 17, 1883.

39 California State Grange, *Proceedings*, 1890, p. 99; 1891, pp. 96–100.

40 Michigan State Grange, *Proceedings*, 1891, p. 53; Kansas State Grange, *Proceedings*, 1889, p. 31.

several subordinate chapters across the state, and New Yorkers praised the law establishing institutes in their state. On the other hand, there were some complaints lodged against extension programs. In 1899 the Illinois State Grange was not altogether contented with institutes being held in that state. It complained that many had been "held in connection with street fairs and other detracting shows and performances not germane to farming" and therefore it recommended that all future extension programs be limited to purely agricultural activities.[41]

As grangers became interested in scientific agriculture, they demanded that knowledge available for farmers' use be expanded through establishment of experiment stations. As a result, the Grange initiated various campaigns to exert pressure on those groups and agencies responsible for improving agricultural research. These included legislatures, school boards, U.S. and state departments of agriculture, and college administrations and faculties.

In several states granges pushed for establishment and improvement of state experiment stations. The bill creating the Ohio Experiment Station was submitted and supported by Senator Joseph H. Brigham, master of the Ohio State Grange. When established, this particular station was governed by an independent control board, and grangers were prominent among its members. Grangers were active elsewhere, too. In New York a special state grange committee investigated Cornell University in 1877. Visitors were impressed by the university's agricultural experimentation program but were disappointed that the school did not make its research findings available to the state's farmers. Because of this dissatisfaction, a special bill providing for an additional appropriation to the university was introduced in the state legislature. The extra money was to be used for sponsoring additional experimentation, for constructing a research center, and for publishing results. On June 26, 1880, the bill became law, and once again members of a state grange were represented on an independent governing board.[42]

The New York grange campaign was unique in several respects. When the state body first began its agitation for a better agricultural

---

[41] Maryland State Grange, *Proceedings*, 1891, p. 6; New York State Grange, *Proceedings*, 1895, p. 91; Illinois State Grange, *Proceedings*, 1899, p. 37.

[42] Alfred C. True, "A History of Agricultural Experimentation and Research in the United States, 1607–1925," United States Department of Agriculture, *Miscellaneous Publication* 251 (Washington, 1937), 94, 97, 98–99; Association of American Agricultural Colleges and Experiment Stations, *Proceedings*, 1889, pp. 41–42.

experimentation service, patrons wanted the state's new station to be part of Cornell's college of agriculture, but later the order advocated something different. By 1878 grangers had concluded that "the proposed can be established more profitably to all concerned, in connection with Cornell University than elsewhere." Three years later, however, New York patrons reversed themselves; now, they wanted research facilities to be separated from the university. They maintained that such an arrangement could serve more farmers; moreover, two distant sites would undoubtedly increase accuracy of experimentation because of the presence of different soil and climatic conditions. Grange arguments were so convincing that in 1882 the state decided to establish a separate center at Geneva.[43]

In their drives for improved agricultural research, Patrons of Husbandry met resistance, quite often from stubborn land-grant college officials who admitted to seeing no value in scientific farming. Master S. C. Carr of Wisconsin charged in 1882 that the university's administration conspired with literary departments, and he claimed that school of agriculture money was being used for classical studies. To eliminate this fraud, Carr proposed that the state establish an independent, permanently financed experiment station. The state grange concluded that the "state and not individual effort should conduct experiments that so materially advance the farming interests." The state legislature, agreeing with the grange proposal, subsequently appropriated money for a state experiment station.[44]

Similar situations developed elsewhere, but grange-proposed solutions were not always as practical. Such was the case in Vermont. In 1885 Overseer Alpha Messer concluded that the University of Ver-

[43] Gould P. Colman, *Education & Agriculture—A History of the College of Agriculture at Cornell University* (Ithaca, N.Y., 1963), 70–71, 81–82; New York State Grange, *Proceedings*, 1881, p. 86; Leonard L. Allen, *History of New York State Grange* (Watertown, New York, 1934), 138.

[44] Vernon Carstensen, "The Genesis of an Agricultural Experiment Station," *Agricultural History*, XXXIV (January, 1960), 18. Charles E. Rosenberg presented another side to farmer interest in experiment stations. He saw it as annoying at times for station staffs. Inquisitive farmers and journalists would wonder why profits were not always shown by "model farms," and they would question as well the sight of dwarfed plants. For them, only healthy-looking specimens should appear, and existence of limp, weak growths was only evidence of failure. Antagonism was also directed at station masters with scientific educations because farmers saw experience on farms as the only necessary qualification for employment. Rosenberg did not specifically link grangers to these types of troublesome activity, but no doubt there was occasional member involvement. Rosenberg, "Science, Technology, and Economic Growth," 2–3.

mont had wasted its land-grant act appropriation and therefore must compensate the state's farmers. One way in which the school could do so, noted Messer, was for it to set aside some of its own funds and establish an independent experiment station.[45]

Most states made little or no progress in establishing experiment stations. On May 8, 1882, Representative C. C. Carpenter of Iowa introduced a bill requesting special federal appropriations for an agricultural experiment station program to be administered by each state. While the bill was awaiting final committee action, agrarian leaders representing land-grant institutions, existing experiment stations, and farmers' groups convened in Washington for meetings. Known as the Washington Agricultural Convention of 1883, this body of delegates chose a special committee to study the Carpenter bill's provisions. Upon completing its investigation, the group enthusiastically endorsed the pending legislation and sent a copy of its report to the National Grange. After carefully studying the Agricultural Convention's special evaluation, the Order of Patrons of Husbandry placed itself on record in favor of the Carpenter bill.[46]

Unfortunately for farmers, Congress adjourned without taking any action on the experiment station bill. When it reconvened, Representative A. J. Holmes resubmitted the measure on December 10, 1883. The Holmes bill, or H. R. 447, was sent to the House Committee on Agriculture. While H. R. 447 awaited final committee action, the Grange launched a systematic campaign. It sent resolutions demanding passage of the experiment station bill to several legislators. Four were read from the floor. Meanwhile, the bill was revised and reevaluated. Chairman Shelby M. Cullom of the House Committee on Agriculture modified the Holmes measure and read his committee's favorable report on the bill to the House of Representatives. In November, 1885, the National Grange met and carefully studied Cullom's revisions. Finally, delegate endorsement came in a special resolution "approving the object and purpose of the bill and . . . [asking for] its passage, so modified in its working details, as to suit and subserve the varied situations and interests in the various States." Despite the Grange's enthusiastic support, the Cullom bill did not receive congressional approval.[47]

---

45 Horton, *History of the Grange in Vermont*, 35; Hills's Marginal Comments.
46 True, "A History of Agricultural Experimentation," 120–22.
47 *Ibid.*, 123; United States *Congressional Record*, 48th Cong., 1st Sess., 1285, 1336, 2034, 2869, and 3076.

When Congress reconvened late in 1885, the experiment station bill had to go through the legislative mill again. This time, however, the scheme had gained two important supporters. Mississippi Senator James Z. George and Missouri Representative William H. Hatch became the new congressional sponsors of the proposal. George submitted his bill on December 10, 1885, and Hatch introduced his on January 7, 1886. Although almost four years had elapsed since Representative Carpenter submitted the original experiment station bill, grangers had not abandoned hope. On the contrary, they continuously sought to improve provisions of the bill. For instance, when the Hatch-George plan came to delegates' attention at the National Grange meeting of 1886, they immediately began to examine it.[48]

With careful scrutiny, the National Grange was able to uncover several flaws and weaknesses in the bill and to suggest corrections. With battles resulting from the Morrill Act still fresh in their minds, committee members suggested an amendment designed to prevent difficulties from arising again. It specifically outlined how revenue from the act should be spent. In states where experiment stations already flourished, money would be given to existing stations and would not be used for establishing new research facilities. On the other hand, in states where recipients of land-grant funds had violated "purposes of that act by neglecting agricultural education" and where no experiment stations had been begun, Hatch-George revenues would go to "duly organized Board[s] of Agriculture" which would use the money for establishment of independent research laboratories "without the intervention of authority exercised by any College, Board, or Faculty." The National Grange forwarded a copy of its proposed amendment to Senator John Sherman, the president protempore of the Senate, who responded by reading the entire grange report on the Hatch-George experiment station bill from the Senate floor. Sherman's action was coupled with readings of other grange petitions, resolutions, and memorials by congressional friends of the order. In fact, evidence of grange agitation for enactment of the Hatch-George bill was apparent in the halls of Congress until March 2, 1887, when it became law.[49]

48 True, "A History of Agricultural Experimentation," 123–27. By this time, Hatch had become the chairman of the House Committee on Agriculture, and George had become the ranking Democrat on the Senate Committee on Agriculture and Forestry. National Grange, *Proceedings*, 1886, pp. 140–41.

49 National Grange, *Proceedings*, 1886, pp. 140–41. A similar amendment was adopted by the next session of the National Grange. See 1887, p. 178. United States

There has been considerable disagreement among laymen and scholars as to whether the Grange was responsible or not for passage of the Hatch Act. Grangers writing about their organization have always given the order full credit for pressuring Congress into passing the experiment station bill, while historians have been less generous. The latter have been more cautious in attributing too much to grange efforts in behalf of the Hatch Act.[50] In reality, it would be very difficult to ascertain the exact role of the order. But it is certain that the Grange made its position known, and no doubt it had some positive effect since the society was considered to be a leading spokesman for the American farmer.

Passage of the Hatch Act did not signal the end of granger activity in behalf of experiment stations. In South Carolina, ten days after the bill had become law, Pomona Grange No. 7 met with McAbee Academy Grange and adopted a resolution asking the state legislature to locate its experiment station in Spartanburg County and pledging to cooperate with all groups willing to work for this objective. Moreover, the New York State Grange annually dispatched teams of inspectors to the state experiment station at Geneva. It was their responsibility to examine the grounds and facilities and to ascertain what progress was being made. Station staff members occasionally used these opportunities to acquaint their visitors with alleged handicaps under which they were working. In September, 1894, Director Peter Collier complained that New York did not have enough research centers. In pleading his case, Collier pointed out that in terms of acreage and value of staple crops, New York exceeded the combined quantities of all of the New England and middle states except Pennsylvania, yet these states had nine experiment stations while New York had only one.[51]

In several states, patrons also worked to establish state departments of agriculture because they wanted governmental agencies which

Congressional Record, 49th Cong., 2nd Sess., 631; True, "A History of Agricultural Experimentation," 128, 129; United States Congressional Record, 49th Cong., 2nd Sess., 386, 420, 631, 694, 746, 870, 1175, 1265; Stemmons, Connecticut Agricultural College, 217.

50 Robinson, The Grange, 1867–1967, 37–38; Mortimer Whitehead, "The Grange in Politics," American Journal of Politics, I (August, 1892), 116; Buck, The Granger Movement, 122; True, "A History of Agricultural Experimentation," 120–29.

51 Greenville (S.C.) Cotton Plant, April, 1887; New York State Grange, Proceedings, 1890, pp. 104–107; 1895, p. 94; For an overview history of experiment stations and their progress through the years, examine: H. C. Knoblauch et al, State Agricultural Experiment Stations: A History of Research Policy and Procedure, USDA Misc. Pub. 904 (Washington, 1962).

could serve and protect them. In North Carolina, Dr. Columbus Mills, one-time master of the state grange, and Leonidas Lafayette Polk, a newspaperman and grange spokesman, took the lead in this movement. Afterward, when Polk was appointed as the first state commissioner of agriculture, the Grange was one of the groups represented on the department's governing board. West Virginia and Colorado grangers also were active in their respective states. In Colorado, of the eight original members who served on the state board of agriculture in 1877, seven were members of the Grange. A similar pattern developed in West Virginia. Granger T. C. Atkeson wrote the bill that created the state board of agriculture, and in 1891 Governor A. B. Fleming chose this influential farm leader to serve with that West Virginia body.[52]

On the national level, the Grange worked in the eighties to bring new dignity and prestige to the Department of Agriculture and to its Commissioner. This post had not yet achieved cabinet status, so Patrons of Husbandry in the 1880s sought to induce Congress to act. Regularly, grange petitions were sent to congressmen. In 1889 Congress reorganized and strengthened the Department of Agriculture and elevated the office of commissioner to full cabinet status.[53]

Of the supplementary activities offered as parts of the organization's adult education program, agricultural fairs and encampments were often most popular with patrons and their families. As the word implies, encampments were the order's version of a bivouac. Patrons either packed their families into buggies and wagons or boarded them on trains and traveled to the campsite, usually a wooded retreat area owned by the organization or by a prominent member. Once there, campers pitched their tents and prepared to spend a few days roughing it.[54]

These retreats featured a variety of activities. There were lectures on scientific agriculture, speeches by leading political figures and by

52 Armstrong, "The Development of Agricultural Education in North Carolina," 42; Stuart Noblin, "Leonidas Lafayette Polk and the North Carolina Department of Agriculture," *North Carolina Historical Review*, XX (July, 1943), 208, 214; True, "A History of Agricultural Experimentation," 90; National Grange, *Proceedings*, 1877, p. 130; Barns, "The Influence of the West Virginia Grange upon Public Agricultural Education of College Grade," 137; William D. Barns, "The Influence of the West Virginia Grange upon Public Agricultural Education of Less than College Grade, 1873–1914," *West Virginia History*, X (October, 1948), 5.

53 United States *Congressional Record*, 48th Cong., 1st Sess., 1885; 49th Cong., 1st Sess., 386, 420, 631, 1175, 1265; 50th Cong., 1st Sess., 183–84, 284.

54 Spartanburg *Carolina Spartan*, August 10, 1887; Illinois State Grange, *Proceedings*, 1889, pp. 17, 35; 1890, p. 29; Baltimore *Maryland Farmer*, September, 1882.

prominent grange officials, exhibits of prize livestock and poultry, fruits, vegetables, and grains, and displays of the latest farm equipment and machinery. On the lighter side, these festive events included colorful militia drills, dances, baseball games, beauty contests, choral and instrumental concerts, balloon ascensions, parades, athletic tournaments, bicycle races, fireworks, and of course, cornucopias of food.[55]

Two of the largest encampments sponsored by the Grange were held at Williams' Grove, a wooded cove thirteen miles southwest of Harrisburg, Pennsylvania, and at Spartanburg, South Carolina. The Pennsylvania retreat began in 1874 as the Tri-state Granger Picnic and developed into a seven-day encampment which attracted 200,000 spectators and campers annually in the 1880s. The event proved to be so successful that grangers in other parts of the country began imitating it. Learning of the overwhelming popularity of the Williams' Grove spectacle, Master J. N. Lipscomb of the South Carolina State Grange asked representatives of the Georgia, Tennessee, North Carolina, and Alabama state granges at the 1885 session of the National Grange if they would be interested in cosponsoring a similar event for southern farm families. Lipscomb found that his associates agreed that the plan had merit, so they consented to meet again with him to draft final arrangements. The result of their efforts was The Inter-State Farmers' Summer Encampment.[56]

Agricultural fairs in the United States have been associated with American farmlife ever since Elkanah Watson coaxed his neighbors near Pittsfield, Massachusetts, into showing their livestock at the town square in 1811. From this humble beginning, agricultural exhibits grew in stature and in numbers throughout the nineteenth century. For the most part, fairs held during the last three decades of the century can be classified into three groups: those sponsored by county and state agricultural societies and by the Farmers' Alliance, those run by governmental agencies such as state boards of agriculture, and those planned and supervised by subordinate, pomona, and state granges.[57]

[55] Spartanburg *Carolina Spartan*, August 10, 1887, August 15, 1888; Patrons of Husbandry, *Prospectus of the First Annual Inter-State Farmers' Summer Encampment* (Spartanburg, 1887), 3–11; R. M. Thomas to William Saunders, August 2, 1888, in Saunders Collection; Baltimore *Maryland Farmer*, September, 1882.

[56] Thomas to Saunders, August 2, 1888, in Saunders Collection; Spartanburg *Carolina Spartan*, August 10, 1887.

[57] Percy W. Bidwell and John I. Falconer, *History of Agriculture in the Northern United States, 1620–1860* (New York, 1941), 187–88, 317–18; New Hampshire State Grange, *Proceedings*, 1898, p. 8.

Grange agricultural fairs originated in the early 1870s, primarily as adjuncts to the organization's adult education program. In the beginning, there were two main purposes for holding fairs; they were supposed to instill a competitive spirit in rural residents, and they were counted upon to draw farmers together to learn of latest advances in agricultural technology, machinery, and scientific farming. Thus, the description given by a Hoosier of his pomona grange's annual fair fits most of these early events: "It is not a horse race, nor a gambling resort, but truly an agricultural fair." [58]

Interest in fairs dropped sharply in most communities after a few had been held in succession, so sponsors either began deviating more and more from original purposes, or granges quit sponsoring them altogether. In 1883 the National Grange admitted that many patrons' fair associations were finding it necessary to introduce "horse races and numerous side shows" to get farmers interested enough to attend fairs.[59]

The editor of a leading southern agricultural journal reported that he had visited the 1878 Alabama State Grange Fair and had left with mixed emotions. After praising exhibits, he proceeded to rap leaders of the order for spoiling their fair by allowing horse races: "If the State Grange of Alabama wish to hold a Fair and secure the confidence and assistance of the moral and religion [sic] portion of the State in the enterprise ... they should bestow less prominence upon horse-racing, which ... breeds a spirit of gambling in the ... young men of the country, and virtually undoes much good that a well conducted Fair would otherwise accomplish." The editor then questioned whether it was "an Exhibition where the father would take pride in taking his son, or the pious mother he[r] daughter?" He concluded his remarks by accusing Alabama grangers of "bankrupting the morals of ... [their] children and bringing agriculture and the Grange itself into disrepute." [60]

Although most grange fairs were local events administered by subordinate and pomona chapters, there were at least three states which had state grange exhibitions. Parent organizations in Alabama, New Hampshire, and Texas each sponsored state agricultural fairs in the nineteenth century. These affairs were managed like most other state fairs at the time. As with other similar events, gate receipts and en-

58 Indianapolis *Indiana Farmer*, February 12, 1881.
59 National Grange, *Proceedings*, 1883, p. 24.
60 Starkville *Southern Live-Stock Journal*, December 14, 1878.

trance fees paid for premiums given to individuals showing the best entries in each class. Similarly, as state fair commissions operated these events for state boards of agriculture, so did special independent associations administer state fairs for state granges. In New Hampshire, this special body did such a commendable job in the 1890s that state grange fairs netted sizable profits.[61]

[61] Alabama State Grange, *Proceedings*, 1875, 18–19; Texas State Grange, *Annual Exhibit of the Texas State Grange Fair, 1890* (Temple, 1890), 3, 6–14; New Hampshire Grange Fair Association, *Official Premium List and Rules and Regulations of Sixth Annual Exhibition* (n.p., n.d.), 1–43; New Hampshire State Grange, *Proceedings*, 1892, pp. 12–13; 1893, p. 13.

CHAPTER VI

Grange Social Activities

Social intercourse among farm families before the advent of the Grange was limited and unsatisfactory. A farmer's lot in the nineteenth century was noted more for its backbreaking work than for its pleasurable aspects. Admittedly, the monotony and tediousness of rural life were relieved somewhat by such spontaneous activities as husking, barn-raising, and quilting parties and by diversions provided by churches and country schools. Still, they failed to lift morale of rural residents sufficiently to satisfy their needs. Certainly farmers did not have sufficient recreational and social activities to render rural existence as attractive as Jefferson's agrarian disciples maintained it would be. Farmers who toiled laboriously from sunrise to sunset never sensed that their homesteads were comparable to the rural paradises envisaged by the Jeffersonians. Instead of living in a "Garden of Eden," many yeomen surmised that they were in fact residing with the Israelites in Egypt.[1]

Something of the loneliness of farm life was known to have been experienced by frontiersmen, and the record of Oliver Hudson Kelley typified the pattern. He learned his lessons firsthand in rural Minnesota and from his trip for President Johnson through the war-ravaged South.

In his immodest but generally accurate account of the founding of the Grange, Kelley noted these problems and concluded that isolation was one of the most serious difficulties confronting rural America. In other words, Kelley perceptively sensed that farmers were suffering from a form of mental agony brought on by loneliness and recognized that something needed to be done. Supplementing the Minnesotan's understanding of the woes of rural society was a narcissistic self-

1 Lawrence *Spirit of Kansas*, July 15, 1875; Wisconsin State Grange, *Proceedings*, 1879, pp. 20–21. For a novelization of life on a farm in the nineteenth century by a man who grew up on one, see Hamlin Garland, *A Son of the Middle Border* (New York, 1917).

portrait of a man with messiah tendencies, who was resolved to change the world rather than curse it. In retrospect, Kelley no doubt was egotistical, but nevertheless he possessed perceptivity, confidence, idealism, and determination—four invaluable assets for any aspiring reformer. Possession of these characteristics explains in large part why Kelley was willing to work so patiently to establish an organization that would give its members not only a chance to learn but also afford them an opportunity to fraternize with each other.[2]

Grange structure was very conducive to social life. Fraternalism and secrecy tended to promote brotherhood, and the organization's rule barring all but farmers and their wives from membership brought unity to the order. Shrouding ritual in secrecy and basing membership upon vocational status gave members confidence in their grange brothers and sisters. Each patron knew that everyone in the organization was pledged to conceal the same rites from outsiders and that his fellow grangers had a common stake in improving agriculture. Moreover, since everyone in the order was engaged in the same occupation, members generally had the same interests. Thus, common association with the Grange bonded rural families into a cohesive force and facilitated development of many new friendships by bringing together for the first time under favorable conditions men and women of similar backgrounds.[3]

The best evidence of the effectiveness of the Grange as a socializing institution was the statements made in behalf of expired brothers and sisters at subordinate, state, and national sessions by friends of the deceased. These eulogies often attested to the fact that new friendships resulted from membership in the order. A special time was set aside at state and national grange meetings for paying homage to departed brothers and sisters. The tributes paid X. X. Chartters of Virginia and his wife at the 1893 session of the National Grange by those who had known them were typical. C. H. Knott of West Virginia vividly recollected how he had first become acquainted with the Virginians at a National Grange meeting held many years earlier in Sacramento and recalled how he had looked forward to seeing them at each succeeding assemblage. Similar accounts were presented by patrons from Penn-

---

[2] Kelley, *Origin and Progress of the Order,* 1–56.
[3] Pauline Swalm, "The Granges of the Patrons of Husbandry," *Old and New,* VIII (1873), 100.

sylvania, California, New Jersey, New York, Delaware, Maryland, and Michigan.[4]

After the Civil War, order leaders resolved that they could not effect a national brotherhood of farmers until the organization had done its part in easing animosities existing between the North and South. Instead of passively accepting the friction as inevitable, they pledged to work diligently for the elimination of all traces of enmity. In 1872 Grange Deputy D. Wyatt Aiken of South Carolina assured farmers in his state and in North Carolina that the organization was not necessarily an evil because of its "Yankee origin." In response to Bourbons' skepticism about the order's merits, Aiken often asked southern reactionaries whether they would discard their clothes if they discovered that the "garments had been manufactured in 'Nazareth.' " [5]

The National Grange campaign to help southerners and northerners forget the Civil War reached a peak at the 1879 session. D. H. Thing of Maine offered a resolution summarizing the order's policy on sectionalism. He stated that grangers "recognized the rights, civil, political and industrial, of each citizen of this Union; that we have no sympathy with sectional feelings and jealousies, with party animosities, with the revival of past issues for party or personal aggrandizement, with that narrow, selfish, unstatesmanlike statesmanship which will result in creating a solid North or a solid South; with bribery, corruption, intimidation, ballot-box stuffing or bull-dozing, either North or South." Thing's resolution was *"unanimously* adopted by a *rising* vote." [6]

In 1874 an Arkansas newspaperman wrote of the reduction of sectional feeling around Fayetteville after the introduction of the Grange in the community. He stated in an editorial that "no system ever inaugurated among the farmers, has done or is doing so much to remove sectional prejudices and personal animosities, as that of the Patrons of Husbandry." [7]

Although grange efforts in behalf of nationalism have either been

---

4 Samples of eulogies may be found in Bloomington Grange (Minn.) No. 482, Records, in Oliver H. Kelley House Papers; Illinois State Grange, *Proceedings,* 1894, p. 49; Indiana State Grange, *Proceedings,* 1882, pp. 54–55; National Grange, *Proceedings,* 1893, pp. 108–17.

5 Charleston *Rural Carolinian,* June, 1872.

6 National Grange, *Proceedings,* 1879, pp. 87–89.

7 Fayetteville (Ark.) *Democrat,* February 7, 1874.

ignored or overshadowed, they nevertheless were as effective as some which have been more widely publicized. While L. Q. C. Lamar's famous eulogy of Charles Sumner has not been slighted as a significant influence upon the reunification movement after the Civil War, similar speeches made regularly in grange halls about departed patrons from the North and South have not been credited for their part in reuniting the Union. After National Grange Master Putnam Darden's death in 1888, grangers from throughout the country mourned his passing by holding special memorial services in honor of their deceased leader from Mississippi. Since these tributes reached so many individuals in granges all across the nation and since the organization taught brotherhood, one may conclude that the order was an important agency fighting sectional friction.[8]

Some granges made concerted efforts to effect new friendships. Exchange meetings were one of the most popular techniques. Cumberland Grange No. 2 of Rhode Island and Olive Grange No. 189 of Indiana each sponsored annual "Neighbors' Nights." On these occasions, grangers from nearby chapters were invited to participate in joint sessions. Besides exchanging ideas and viewpoints, patrons also met their grange brothers and sisters from other locals.[9]

In view of the fact that granges made determined efforts to produce new friendships, it was ironic that a few chapters acted coolly and hesitantly when dealing with brothers and sisters from locals weakened by ebbing interest in the order. In reality, there was always a possibility of ugly situations developing whenever weak subordinates asked healthier ones to absorb them. Members of strong chapters evidently felt that such consolidations posed threats to the homogeneity and tranquillity of their chapters; at any rate they often tended to shy away from these requests. Occasionally, they showed their disapproval by simply setting merger deadlines. If the grangers wishing acceptance failed to comply by a specified date, they were ordered to be initiated again. If they refused to go through the procedure a second time, they were dropped from the organization.[10]

In an address delivered at the 1879 session of the Wisconsin State Grange, Master H. C. Sherwin discussed the impact of the Grange

[8] Paul H. Buck, *The Road to Reunion: 1865–1900* (New York, 1937); Wisconsin State Grange, *Proceedings*, 1888, pp. 27–29.

[9] Olive Grange, Secretary's Minutes, June 28, 1884; Rhode Island State Grange, *Proceedings*, 1894, p. 7.

[10] Church Hill Grange, Secretary's Minutes, February 26, March 10, 1876.

upon farmers' lives and told of changes resulting from the organization that he had observed across the country. "Thousands of neighborhoods where but little if any sociability existed have become the centres [*sic*] of social enjoyments. Where once we had hardly neighbors we are now brothers and sisters." [11]

With social activities being such an integral part of the grange program, securing an acceptable facility for accommodating order functions became one of the first tasks confronting subordinate chapters upon receipt of their charters. In many communities, existing fraternal bodies—Masons, Good Templars, and Knights of Jericho—cooperated with fledgling granges by extending helping hands. These groups often permitted Patrons of Husbandry to use their halls for nominal fees. In other areas, grangers found temporary quarters in church sanctuaries, schools, and vacant stores; in neighborhoods where no meeting places existed, patrons held sessions at members' homes. This procedure usually was followed in emergency cases, too. For example, after the lodge hall owned by the Fairfield Grange of Huron County, Ohio, burned in 1896, members met at each other's residences.[12]

Most subordinate chapters eagerly awaited the day when they could vacate their temporary quarters and move into their own lodge halls. Realization of this objective took much preparation and sacrifice. To plan and construct a suitable and comfortable edifice for holding grange meetings required a willingness on the part of members to surrender their spare money, time, and energy. Since local membership dues seldom produced treasuries large enough to finance the building and equipping of halls, members commonly found it necessary to dig deeply into their own pockets for additional money. In many local chapters, patrons made pledges to defray property and construction costs. Occasionally granges financed their building projects by other means, such as cultivating a vacant piece of property and donating profits to a building fund. One effort of this type in Kansas produced

11 Wisconsin State Grange, *Proceedings*, 1879, pp. 20–21.

12 Booneville (Ark.) *Enterprise*, July 23, 1875; Raleigh Grange (N.C.) No. 17, Secretary's Minutes, July 19, 26, August 2, 9, 16, 23, September 6, 13, 1873, in Duke University Papers of the Patrons of Husbandry, Duke University Library, Durham; Des Moines *Iowa Homestead and Western Farm Journal*, February 16, 1872; Wilmington Grange, Secretary's Minutes, February 5, 1876; Branchville Grange (Branchville, Ga.) No. 425, Secretary's Minutes, May 6, September 16, 1875, in Branch Family Papers, Southern Historical Collection, University of North Carolina Library, Chapel Hill; Bangor (Me.) *Dirigo Rural*, December 5, 1874; Cleveland *Ohio Farmer*, July 23, 1896.

on a twenty-acre plot of land six hundred bushels of shelled corn that brought a return of more than $200.[13]

In order to prune expenditures, patrons often were asked by lodge hall committees to volunteer their laboring skills. Thus, by putting their members' carpentry, masonry, and decorator talents to use, granges were able to erect and furnish attractive quarters for less than $1,000. Outlays ran much higher for ready-built structures. In the case of Raleigh Grange No. 17 of North Carolina, the Patrons of Husbandry belonging to this local had to assess themselves twenty dollars in order to raise enough funds to buy a $3,500 house. Grange-hall-raisings were similar to barn-raisings. Crews of amateur tradesmen worked enthusiastically while their wives gossiped, prepared dinners, and made coffee for their husbands. Upon completion of rough work, matrons added a woman's touch to interiors of buildings by draping windows, hanging pictures, and putting potted plants in conspicuous locations around the room; male members cultivated and seeded yards and painted walls. Interior fittings came principally from two sources: home workshops and factories. Ingenious patrons produced many necessary pieces of furniture, such as tables and benches, while other appendages like spittoons, organs, upholstered chairs and divans, and lamps were purchased as soon as money became available for them. Upon completion, grange halls were exhibited at public dedication ceremonies. The total number of grange halls built during the nineteenth century is unknown but there were literally thousands. In Indiana, a stronghold of grangerism in the 1870s, approximately five hundred lodge halls had been completed by the summer of 1876.[14]

Granges did not always find it expedient to rely solely upon their own limited resources. Some edifices were products of joint efforts of granges and other groups. Although cooperative enterprises of this nature were not common, they nevertheless worked satisfactorily in

13 Des Moines *Iowa Homestead and Western Farm Journal*, February 16, 1872; Church Hill Grange, Secretary's Minutes, February 14, 1879; Raleigh Grange, Secretary's Minutes, November 15, 1873; Topeka *Kansas Farmer*, January 10, 1877.

14 Des Moines *Iowa Homestead and Western Farm Journal*, February 16, 1872; Church Hill Grange, Secretary's Minutes, December 16, 23, 29, 1873, January 6, 1874, March 10, 1876, May 10, July 26, October 5, November 22, 1878, February 14, May 9, 23, 1879; Raleigh Grange, Secretary's Minutes, October 11, 25, November 15, 1873; Indianapolis *Indiana Farmer*, January 19, 1878, March 25, 1882; Clarke County Grange, Secretary's Minutes, September 2, 6, 1873; Olive Grange, Secretary's Minutes, October 5, 1889; Washington Bowie to Saunders, November 26, 1874, in Saunders Collection; Chicago *Prairie Farmer*, July 1, 1876.

areas where no social and professional gulfs existed between fraternal orders and where poverty prevented different societies from acting independently. Blessed with good rapport and confronted with poor prospects of raising enough money to build three separate facilities, the Patrons of Husbandry, Knights of Jericho, and Masons of Pomaria, South Carolina, united in a single objective and effected an arrangement whereby each organization did its share to finance, construct, and administer quarters housing all three groups. After the initial idea of union had been broached, the orders chose interfraternity committees to study the proposal's feasibility. The three brotherhoods, being well satisfied with their preliminary investigations, pushed ahead and built a two-story structure, thirty feet by sixty. The lodges met on the second floor and rented the bottom level to a businessman who converted it into a dry-goods store. Thus, through shrewd management and wise planning, the three brotherhoods gained a combined annual income of $200 from their joint real-estate holding.[15]

It was not uncommon for grange halls to yield small incomes. In an era predating the general availability of school gymnasiums, auditorium facilities were scarce. When assemblages of people met for an indoor activity, few adequate structures were available in most rural communities. Therefore, many granges surmised that their idle facilities could be turned into rewarding investments. In fact, Minnehaha Grange No. 398 of Richfield, Minnesota, found that rental prospects and demands for its hall were so promising that it formed Edina Hall Association to handle arrangements. This body, financed by sale of stock and governed by its own officers, was chartered by the legislature. Judging by the association's books, the operation was successful; it consistently showed profits for the period 1879–1907. Although few granges engaged in rental activities on a comparable scale, several others did receive small returns from their halls. In grange ledger books, there are numerous examples of outsiders renting lodges at daily rates of around $3.50.[16]

Being a secret organization, the Grange in the conduct of its meetings concealed many activities from the general public, but this did

15 Waukon (Iowa) *Standard*, February 13, 1873; Pomaria Grange (S.C.) No. 27, Secretary's Minutes, June 28, 1872, February 7, 21, 1874, Clemson University Library, Clemson, S.C.

16 Minnehaha Grange (Richfield, Minn.) No. 398, Secretary's Minute Book, 1879–1907, of the Minnehaha Grange (Edina) Hall Association, Minnesota State Historical Society, St. Paul; Oliver H. Kelley House Papers, March 19, 1884.

not mean that during sessions the order completely insulated its membership from outside forces. In many cases, just the opposite was true. Nonmembers frequently were guests at grange assemblages, and they often participated in meetings. On state and national levels of the organization, prominent educators, public officials, and scientists regularly appeared at conclaves and delivered addresses. Participation by such individuals no doubt increased grangers' confidence and made them feel important. Illinois grangers who met Governor John Peter Altgeld at the 1895 session and Pennsylvania farmers who at a state grange meeting had the opportunity to shake hands with Governor James A. Beaver had new and enlarged views of their worth.[17]

Outside contacts were not limited to guest appearances by educators, politicians, and agricultural experts at annual meetings. A spokesman for farmers and their causes in the late nineteenth century, the Grange could not remain aloof from other organizations. Instead it came into contact with a wide spectrum of special-interest groups, ranging from temperance leagues to labor unions. In assessing these associations between granges and other bodies, no generalizations can be made because individual chapters tended to treat interorganizational exchanges differently. In their dealings with other bodies, granges followed neither a regional nor a logical pattern of consistency. There were some underlying reasons behind this fluidity. Grange constitutions made no references to the subject of intergroup relationships, and leaders at state and national levels rarely formulated policies for such confrontations.[18]

No better example illustrating order inconsistency in exchanges with other groups can be cited than the nebulous Grange-Farmers' Alliance relationship of the 1880s and 1890s. In some instances, the

[17] Illinois State Grange, *Proceedings*, 1890, p. 8; 1895, p. 38; 1899, p. 19; New Hampshire State Grange, *Proceedings*, 1880, p. 63; 1891, pp. 74–76; Indiana State Grange, *Proceedings*, 1881, p. 27; 1894, pp. 29–30; Wisconsin State Grange, *Proceedings*, December, 1882, p. 31; 1888, pp. 15–16; Pennsylvania State Grange, *Proceedings*, 1887, pp. 12–13; New York State Grange, *Proceedings*, 1879, pp. 62–66; 1889, pp. 54–75, 80; 1893, pp. 91; 1894, pp. 135–36; 1895, pp. 106–108; Ohio State Grange, *Proceedings*, 1897, pp. 40–45, 52; 1898, pp. 29–30; National Grange, *Proceedings*, 1878, p. 18; Albany *Cultivator and Country Gentleman*, February 7, 1878.

[18] Greene County Pomona Grange, Secretary's Minutes, August 21, 1891; Borrors Corners Grange, Secretary's Minutes, March 26, 1898; Chicago *Prairie Farmer*, April 5, 1873; Massachusetts State Grange, *Proceedings*, 1886, pp. 46–47; 1899, pp. 54–58; Virginia State Grange, *Proceedings*, 1879, p. 14; South Carolina State Grange, *Proceedings*, 1877; National Grange, *Proceedings*, 1886, pp. 118–19; 1890, pp. 80–81; 1895, pp. 48–49; National-American Woman Suffrage Association, *Proceedings*, 1896, p. 151; 1897, p. 68. Information gathered from constitutions listed in the bibliography.

older body cooperated fully with its upstart competitor and praised its endeavors; at other times, open bitterness raged among the two rural societies. When harmony existed, alliancemen and grangers acted like a mutual admiration society. Members of the two organizations crossed lines to cosponsor picnics, attend each other's conventions, and pledge themselves to a common objective—improvement of the nation's agricultural class. But when animosity and rivalry were present, relations between the two agrarian bodies left much to be desired. References were made at grange sessions to the "wildcat schemes" of the Farmers' Alliance, and grange leaders appealed to patrons' "conservatism to denounce these schemes of getting something for nothing." Grangers also showed their general disapproval of their rival's program by spurning invitations from the National Farmers' Alliance to attend its meetings. How much of this can be linked to policy disagreements and how much to resentment and jealousy caused by the sudden success of the Farmers' Alliance will never be known. One can be certain, however, that some grangers waited for the day when former members, who had abandoned the Grange for the more militant Farmers' Alliance, would return to the fold following the collapse of the newer organization.[19]

On the other hand, organizations like the National-American Woman Suffrage Association (NAWSA) and the Woman's Christian Temperance Union (WCTU)—groups that shared grange objectives but did not compete with it for members—enjoyed splendid relations with Patrons of Husbandry. Throughout the 1880s and 1890s, crusading efforts of Frances E. Willard and Susan B. Anthony won enthusiastic endorsements and periodic votes of confidence from reform-minded granges. Although, from its creation in 1867 the order had been committed unequivocally to these ladies' causes, it was not until the Second Grange Movement that patrons actively demanded equal rights for women and prohibition of alcoholic beverages. Pleased, the WCTU and the NAWSA often showed their gratitude by communicating their appreciation to sessions of state and national granges.[20]

19 National Grange, *Proceedings*, 1881, p. 147; 1888, p. 104; 1888, p. 72, 104; 1890, pp. 77–78; 1892, p. 113; Illinois State Grange, *Proceedings*, 1890, p. 77; 1891, p. 37; New York State Grange, *Proceedings*, 1891, p. 116; Augusta *Southern Cultivator and Dixie Farmer*, December, 1888; Minnehaha Grange, Secretary's Minutes, May 17, 1884; Social Grange, Secretary's Minutes, September 16, 1891; New Hampshire State Grange, *Proceedings*, 1890, p. 48; Smith, "The Grange Movement in Texas," 306–307.

20 National-American Woman Suffrage Association, *Proceedings*, 1896, p. 151; 1897, p. 68; Illinois State Grange, *Proceedings*, 1889, p. 64; National Grange, *Proceedings*, 1890, pp. 80–81; 1895, pp. 48–49.

Besides giving members chances to meet brothers and sisters from other areas, state and national sessions afforded other opportunities. For one thing, they released rural Americans from their restricted home environments and opened new vistas for them. Each year state and national granges pigeonholed a certain percentage of their annual budgets for mileage and per diem expenses. These allotments enabled patrons selected to attend state and national assemblages, to go from coast to coast for meetings, and to be remunerated for their expenses.[21]

Choosing the right convention site was important to grangers, so they looked at many factors. Foremost in their minds was availability of auditorium space. Another important consideration was hotel accommodations. At times, accessibility of towns also entered the picture as did the number of side attractions offered by communities. Occasionally patrons merely solicited bids and then coldly pitted advantages of one place against those of another. The city endowed with the most facilities at the least amount of money then was selected. As a matter of fact, this procedure often was used to designate inns as official headquarters. If everything else were equal, hotels with lowest room rates were dubbed seats of conventions. Of course in areas where the order was strong, vying among innkeepers for this honor was very spirited since it was not unusual for several hundred delegates to attend these conclaves. As every hotel proprietor knew, filled rooms brought in enough revenue to cover fixed costs and guaranteed a profit.[22]

With a group as large as the Grange, it was only natural that members were not always satisfied with the sites selected for annual meetings. In 1876 Oliver Kelley fretted because the executive committee of the National Grange had bypassed Louisville, his choice, in favor of Chicago. He thought that the decision was unwise because the order had spent thousands of dollars to secure a permanent office in the Kentucky city and because Philadelphia was a better choice than the Windy City. The Pennsylvania metropolis was the setting for the nation's centennial celebration and exhibition, a noteworthy attraction of considerable merit.[23]

[21] Information gathered from grange proceedings listed in the bibliography.

[22] Vermont State Grange, *Proceedings*, 1875, p. 23; Indiana State Grange, *Proceedings*, 1889, p. 40; New York State Grange, *Proceedings*, 1892, p. 91; 1895, p. 92; Atlanta *Rural Southerner*, July, 1874; Nebraska State Grange, Secretary's [handwritten] Journal of Proceedings, December, 1887.

[23] Indiana State Grange, *Proceedings*, 1883, pp. 36–37; Kelley to Ellis, July 15, 1876, in Taber Papers.

Of course, the quantity of money available for convention purposes fluctuated in direct proportion to total membership. When more individuals paid dues, larger sums gathered in travel chests. Accumulating surpluses occasionally proved to be sources of temptation. Like Adam being pulled by Satan to partake of the "forbidden fruit," delegates often succumbed to the seductive voice of the dollar. Availability of treasury reserves led them to vote themselves extended vacations at order expense. Obviously, the 62 "official" representatives who journeyed to Sacramento, California, in 1889 for a National Grange session and the officers who planned this junket were guilty of extravagance. The "vacationers" generously dipped into treasury funds to cover their heavy travel outlays; in all, they bestowed upon themselves reimbursements averaging approximately $181 per man.[24]

Travel opportunities to conventions and social contacts were not the only ways in which state and national sessions broadened delegates' horizons. Side activities also brought many intangible rewards, such as enduring remembrances of places visited in connection with order conventions. As the records show, these affairs were not all work and no play, and serious grange business did not take all of the delegates' time and energy. Consequently, grange convention-goers customarily paused at least once at each annual meeting to take a group sightseeing excursion. Common attractions included governors' mansions, libraries, penitentiaries, college campuses, orphanages, national shrines, museums, municipal buildings, academies, sanitariums, and riverside facilities. For many convention veterans, a climax was reached in 1880, when Rutherford B. Hayes personally invited grangers to visit him at the White House. Moving through hallowed halls and rooms of this famous edifice, with the president himself as a private guide, no doubt more than one horny-handed farmer imagined that he felt the presence of Monroe, Jackson, and Lincoln. Certainly no reminders were necessary to tell these fortunate grangers that illustrious men once strolled the same corridors and inhabited the same chambers.[25]

24 New York State Grange, *Proceedings*, 1892, p. 93; Indiana State Grange, *Proceedings*, 1881, pp. 33–34; Wisconsin State Grange, *Proceedings*, 1879, pp. 74–76; 1880, pp. 79–80; National Grange, *Proceedings*, 1889, pp. 113–14.

25 National Grange, *Proceedings*, 1880, p. 39; 1889, pp. 5–6; 1890, pp. 70–71; 1895, pp. 215–17; North Carolina State Grange, *Proceedings*, 1878, p. 18; 1882, n.p.; Missouri State Grange, *Proceedings*, 1889; New Hampshire State Grange, *Proceedings*, 1894, p. 4; Massachusetts State Grange, *Proceedings*, 1898, pp. 7, 9; *Charleston Rural Carolinian*, March, 1875; Indiana State Grange, *Proceedings*, 1894, p. 37; Wisconsin State Grange, *Proceedings*, 1883, pp. 8–9; 1884, p. 8; 1888, pp. 9, 23.

In 1888 the Topeka *Capital-Commonwealth* told of another interesting excursion trip taken by National Grange delegates. According
to the article, conventioners rose with the break of dawn, boarded
railroad coaches at Topeka, and then disembarked later for a day at
Kansas Agricultural College in Manhattan. The reporter covering the
venture noted in his account how he had been impressed by the wide-
eyed farmers and their wives who gazed steadily through their windows in order to catch every sight as the train slithered briskly down
the track. Upon arriving at their destination, the travelers were met
and welcomed at the platform by a host of dignitaries, including the
chairman of the community board of trade and the president of the
college. After being fed "an elegant and substantial meal" prepared
by the ladies of the local relief corps, the 150 guests climbed into
buggies and carriages and rode to the campus, where they spent "two
hours . . . inspecting the varied details of farm and college, and then
it was time to get to the train for the return trip." It was during this
short drive to the train depot that the "only accident which marred
the day's enjoyment occurred." As a result of a hard jolt incurred in
passing over a rough crossing, a rear seat of one of the buggies jarred
loose from its moorings and hurled its two occupants to the hard
pavement. The two passengers were fortunate to have sustained only
minor injuries because they both landed on their heads and shoulders.
But even this freak accident did not deter the sightseers from agreeing
upon their return to Topeka that their "day [at Manhattan] could
hardly have been made more pleasant than it was." [26]

Patrons took fewer side trips during the early years of grangerism.
Delegates were more idealistic then; members spurned frivolity and
devoted their time to sober issues facing farmers and their families in
rural America. Therefore, it was not uncommon during the peak
years of the First Granger Movement for conventioners to decline invitations and concern themselves with serious matters related to farm
life.[27]

No doubt renewing old acquaintanceships and taking group tours
meant much to delegates attending grange conventions, but there
were other activities which enlivened grange sessions. For many patrons, the music and poetry wedged between dry committee reports

[26] Topeka *Capital-Commonwealth*, November 18, 1888.
[27] Oregon State Grange, *Proceedings*, 1877, p. 26; Wisconsin State Grange, *Proceedings*, January, 1874, pp. 17–18; Jeremiah Henry to Byron V. Henry, February
23, 1874, in Jeremiah and Byron V. Henry Papers, Duke University Library, Durham, N.C.

and votes on resolutions brought true enjoyment.[28] Singing familiar
stanzas to order favorites like "Rally Round the Grange," "Sow Thy
Seed," "Sowing and Reaping," and the "Patrons' Harvest Song" not
only prepared members emotionally for more routine business matters ahead, but also lifted their spirits by conveying messages of hope
and relief from the burdens of the day and by making farmers feel
that their calling was the noblest of all. For example, the grangers'
adaptation of "Rally Round the Flag" certainly projected gusto and
steadfastness:

> We will rally round the Grange,
> We will rally once again,
> Shouting the Farmer's cry of Freedom.
>
> We will rally to the Grange, our rights to maintain,
> Shouting the Farmer's cry of Freedom.
> The Patrons forever, hurrah, then hurrah!
>
> Down with th' oppressor, up with our star,
> We will rally to the Grange, our rights to maintain,
> Shouting the Farmer's cry of Freedom.[29]

Occasionally, guest "artists" sang and played musical instruments. At
the 1898 session of the Massachusetts State Grange, Rosalie Morse of
Southbridge presented a violin solo, and at a meeting of the California State Grange, Granger James G. Clark sang "We Have Drunk
from the Same Canteen" and followed it up with an original composition, "Star of My Soul." [30]

Music also played a very vital part in the programs of most subordinate granges. Secretaries' minutes for some locals abound with references to pleasures gained from singing about the Grange and about
the joys of being a member of the agricultural class. Nor was it unusual for granges to have their own choirs and orchestras. These en-
sembles performed at local meetings and, on occasion, at state and
national sessions. With so much emphasis being placed on music at
grange functions, it was no wonder that pump organs became very
popular fixtures in grange halls.[31]

[28] Information based on surveys of state and national grange proceedings; see
National Grange, *Proceedings*, 1899; New York State Grange, *Proceedings*, 1892,
p. 95; Massachusetts State Grange, *Proceedings*, 1898, p. 84; California State
Grange, *Proceedings*, 1890, p. 56.

[29] National Grange, *Songs of the Grange* (Philadelphia, 1874).

[30] Massachusetts State Grange, *Proceedings*, 1898, p. 84; California State Grange,
Proceedings, 1890, p. 56.

[31] Social Grange, Secretary's Minutes, 1891–99; Borrors Corners Grange, Secretary's Minutes, March 4, April 15, 29, 1876; New York State Grange, *Proceedings*,

During the first years of the order's existence many members consistently misunderstood the Grange's mission. As they interpreted it, singing songs and taking excursions were not legitimate granger activities; the organization's prime purpose was unifying farmers against their alleged "oppressors." With this conception in mind, many patrons naïvely stamped out the true bone and sinew from the order's program—giving rural residents another social outlet and affording them an opportunity to learn from neighbors. Failing to grasp the order's primary objectives resulted in the passage of resolutions barring all "side shows and public amusements" from lodge halls. In most granges, however, initial stigmas of seriousness and total commitment eroded completely after a year or two of organization. Paralleling this movement was a general relaxation of chapter policies. Restraints and stuffiness thus gave way to expressions of social consciousness and enjoyment, and as these characteristics became more apparent at local sessions, patrons celebrated more regularly. These changes in attitude about what constituted a good grange meeting naturally prompted grangers to add more social features to their meetings. With this development, sessions held for specific purposes and without particular social connotations suddenly evolved into festive affairs of considerable magnitude.[32]

One event which certainly moved in this direction was the installation service. Beginning in the early 1870s as a drab ceremony for swearing in recently elected officers, it had already developed into an annual social spectacle by the middle of the decade, reserved for the first regular meeting of each year and "open[ed]" generally to members and their friends. As a rule, the first part of the grange installation program was sedate and solemn and featured the ceremonious ordeal of oath administrations. However, with the conclusion of this portion of the session, the mood changed abruptly from one of reverence to one of gayety. In an atmosphere not unlike that existing at governmental centers on inauguration or coronation days, members and their guests witnessed pledge-taking rituals and then relaxed by dancing and eating stately dinners. Evidently, most who attended these installation balls enjoyed themselves. The affairs became a custom which spread throughout the grange belt, and contemporary ac-

---

1892, p. 95; Indiana State Grange, *Proceedings*, 1894, p. 43; Printed Grange Circular entitled *Strictly Confidential*, in Joseph H. Osborn Papers, Oshkosh Public Museum, Oshkosh, Wisc.; Ohio State Grange, *Proceedings*, 1876, pp. 33–34.

[32] Borrors Corners Grange, Secretary's Minutes, November 6, 1874. Compare with minutes taken at later sessions.

counts glow with reports of participants who found them to be pleasant events. One eyewitness of a Nebraska installation reported that his brothers and sisters and their guests shuffled their feet to rhythms of a dance band until "bed-time" and then observed that "all went home feeling happy." [33]

There were at least four other specific occasions during the year when many granges convened primarily for social purposes. Many local chapters began to celebrate the completion of another year's crop season with special observances known as harvest days. Although these autumn events varied from place to place, they nevertheless possessed some common characteristics. Seen at all harvest days were overflowing baskets of delicious food, people milling about and enjoying themselves, and symbols of the year's production, such as shocks of wheat, stalks and ears of corn, and clusters of fruits and flowers. These samples were supposed to represent the total labor expended by members to plant and gather their crops and the provisions given them by Mother Nature. Thus, harvest days tended to blend feelings of thankfulness to God with those of natural joy and relief farmers felt as they anticipated the slack winter months immediately ahead.[34]

Similarly, many chapters specified one day each year for honoring children. Rural family ties were very close in the nineteenth century partly because parents needed their sons' and daughters' assistance to complete farm tasks. But the idea of children doing chores around the barnyard and helping with the harvesting of crops had become so engrained in rural parents' minds by the time the Grange appeared that mothers and fathers often forgot what blessings their young ones really were. Grange leaders, however, set out to correct that situation. Local granges began hosting parties for members' children. Known as grange children's days, these events included games, favors, songs, and as always—"good things" to eat.[35]

[33] San Francisco *Pacific Rural Press*, February 17, 1877; Center Grange (Yankee Hill, Nebr.) No. 35, Secretary's Minutes, January 15, 1876, Center Grange Records, Nebraska State Historical Society, Lincoln; Olive Grange, Secretary's Minutes, December 25, 1886; Denver *Rocky Mountain Weekly News*, March 18, 1874; Valley Grange (Gardner, Kans.) No. 312, Secretary's Minutes, December 13, 27, 1873, January 10, 1874, Valley Grange Minutes, University of Kansas Library, Lawrence; Indianapolis *Indiana Farmer*, January 19, 26, 1878; Social Grange, Secretary's Minutes, January 1, 1894; Greene County Pomona Grange, Secretary's Minutes, May 22, 1887; Lincoln *Nebraska Patron*, March 3, 1875.

[34] Chicago *Industrial Age*, August 20, 1873; Amherst (N.H.) *Farmers' Cabinet*, April 9, 1878; Big Springs Grange, Secretary's Minutes, September 25, 1874.

[35] Cleveland *Ohio Farmer*, July 29, 1897; Rhode Island State Grange, *Proceedings*, 1894, pp. 15–16.

Dates commemorating the birth of the order and of the Declaration of Independence were two other occasions celebrated with a social twist by Patrons of Husbandry. To mark the founding of the organization members often set aside December 4, the official anniversary of the Grange's establishment, for "sumptuous banquet[s]" and dances. Most of these events were probably similar to one held in 1874 at Liberty School House in West Fork, Arkansas. At this particular affair, brothers and sisters of Liberty, West Fork, and Crawford gathered and partook of a buffet-style dinner arranged on a long table, sixty feet by three. "There were nice boiled hams, turkeys, chickens, and a variety of meats, cakes of all kinds, and a general supply of 'knick-knacks' and everything that [the] heart could wish." One participant recalled that there had been an "abundance for all and plenty left." According to him, the anniversary dinner had been so enjoyable that it would "be long remembered by those" who had been present "as a day pleasantly and profitably spent." [36]

No social attractions sponsored by granges surpassed in popularity and splendor the order's Fourth-of-July "pic-nics." In communities where the organization was strong, the pattern was the same. On Independence Day mornings, patrons gathered their children, packed their wagons, and proceeded to a central meeting place. From there, in step to music provided by local bands and beneath flying banners and flags, grangers with their families marched or rode in unison to a favorite grove. There, under the shade of trees, grange Fourth-of-July celebrations as a rule featured a multitude of activities, including music, games, contests, dances, fireworks, basket lunches, and speeches by politicians and prominent order officials. The latter were almost always at a premium on Independence Days because there were more picnics than there were state and national officers of note.[37]

Holidays and special occasions were not the only times when grangers had "jolly good time[s]." Throughout the year, granges sponsored picnics and dinners. In fact, during the summer months, patrons tried to meet outdoors as often as their busy summer-work

[36] Denver *Colorado Farmer*, January 1, December 17, 1885; West Butler Grange, Secretary's Minutes, November 23, 1875; North Star Grange, Secretary's Minutes, December 5, 1868; Fayetteville (Ark.) *Democrat*, December 12, 1874.
[37] Chicago *Tribune*, July 7, 1873; Denver *Colorado Farmer*, July 10, 1884; Macon *Missouri Granger*, June 22, 1875; J. A. Swain to Osborn, June 14, 1875, Thomas Wilcock to Osborn, June 21, 1875, both in Osborn Papers; John Rehrig to Wood, June 15, 1874, in Wood Papers; Cottage Grange (Grovewood, S.C.), Special Announcement in the Secretary's Minute Book, n.d., Cottage Grange Records, Duke University Library, Durham, N.C.

schedules permitted and as frequently as the weather allowed them to enjoy comfortably basket dinners, strawberry festivals, ice cream socials, barbecues, and beach parties. As a result, these events seemed to spring up like Topsy everywhere the order existed. For the most part, they were spontaneous and without direction from state or national granges; few, if any, comprehensive blueprints were ever laid out by parent bodies for these types of activities.[38]

This did not mean, however, that granges never developed their own regular routines. In fact, the opposite often happened. For example, members of one particular local in Ohio alternated throughout the summer months by taking turns at hosting sessions in their yards. Each week, patrons and their families carried baskets laden "with the substantials and luxuries of life" to different homes. Naturally, much more emphasis was placed upon social values, informality, and enjoyment at these outdoor meetings than was the rule at winter sessions. But still, lively discussions ensued at these fresh-air conclaves, and the lecture hour was not abandoned completely. While several granges preferred intimate little affairs like those held by the Ohio chapter, other locals enjoyed combining with neighboring groups to plan mammoth picnics. In areas of unusual grange strength, outdoor gatherings drew well in excess of five thousand participants and attracted considerable attention. On appointed days entire communities emptied into large groves for afternoons of basket lunches, speeches, games, and music; and correspondingly, all other business was suspended because these events became "the absorbing theme[s]" of the moment.[39]

One effect of large attendance at grange outings was not altogether a blessing. As picnics attracted bigger crowds and more attention, more and more hucksters and budding politicos gravitated to these events. Within a short time, they were appearing in such large numbers that they were a nuisance. As a result, grangers began clamoring for the exclusion of contentious elements from all order functions. However, little came from these protests because most patrons en-

---

38 Massachusetts State Grange, *Proceedings*, 1884, pp. 8–9; Social Grange, Secretary's Minutes, May 16, 1894; Nashville *Rural Sun*, August 7, 1873; Picnic Committee of Shocco Grange (Warrenton, N.C.) No. 301 to Polk, June 28, 1878, in Polk Papers; Cottage Grange, Secretary's Minutes, June 11, 1875.

39 Cleveland *Ohio Farmer*, October 13, 1877; The Reverend Samuel Andrew Agnew, Diary, August 12, 1875, in Southern Historical Collection, University of North Carolina Library, Chapel Hill; H. C. Deming to Saunders, August 11, 1874, in Saunders Collection; Charleston *Rural Carolinian*, October, 1874.

joyed seeing men like Cyrus C. Carpenter, the Governor of Iowa, address their outdoor gatherings. Moreover, they recognized that it would be an exercise in futility to bar undesirable politicians and admit popular individuals like Carpenter; it also would be a clear violation of the order's rule against partisan politics.[40]

The first frost usually signaled the end of the Grange's summer schedule of activities, but the order's social program did not go into hibernation until the following spring. On the contrary, grangers simply confined their social events to their lodge halls and substituted dinners for picnics as the principal bills of fare. Oyster suppers and "bounteous" dinners seemed to be the most popular of all the winter social activities planned by granges. Many members also found dancing to be a most pleasing recreational and social diversion at grange sessions. According to one account, "mock, hypocritical dignity was thrown to the winds, and the gay company, like a band of loving brothers and sisters of the same parents, reveled in the mazy dance like so many, pure-hearted, innocent children." [41]

In any organization as large as the Grange, there were bound to be some members who prudishly disapproved of almost everything. This was especially true in the late nineteenth century, an era of piety and crusades against alcoholic beverages. Reflecting the mood of the times, one sincere Minnesotan sternly warned his brothers and sisters in the Grange that if balls were not banned immediately from granges, they would "kill the order in one year." To this individual and to many others like him, dancing was "the flowery entrance-gate to hell." [42]

During the cold winter and spring months, some granges found that holding sessions at different members' homes proved enjoyable. Riverside Grange No. 37 of Virginia followed this formula and discovered that "great interest ... [was] shown in ... [these] meetings and ... [that the] suppers [preceding these neighborly sessions] add

[40] New York *Times*, June 12, 1873; Macon *Missouri Granger*, October 5, 1875; Agnew Diary, August 12, 1875, in Southern Historical Collection.

[41] Portsmouth *Virginia Granger*, February 22, 1883; Social Grange, Secretary's Minutes, December 16, 1891; Douglas County Pomona Grange, Secretary's Minutes, February 26, 1876; Clarke County Grange, Secretary's Minutes, December 17, 1873, January 14, April 1, 1874; Indianapolis *Indiana Farmer*, February 20, 1875; Sarah G. Baird, Diary, January 4, 1896, in Sarah G. Baird Papers, Minnesota State Historical Society, St. Paul; Garnett (Kans.) *Weekly Journal*, n.d., Hanway Scrapbook, V, 3; Olive Grange, Secretary's Minutes, October 4, November 12, 1887; Topeka *Kansas Farmer*, February 5, 1879; North Star Grange, Secretary's Minutes, December 16, 1868.

[42] St. Paul *Grange Advance*, May 27, 1874.

[ed] much to the enjoyment of the occasion." Similarly but deviating somewhat from this plan, a grange in Adrian, Michigan, held only special functions at the residences of its members and netted very favorable results. Two of the more pleasant social experiments attempted by this particular chapter in the spring of 1876 were a maple-sugar social hosted by one brother and attended by more than seventy persons and a tree-planting party and supper held at the farm of another.[43]

Naturally, with social diversions assuming such an important part in the Grange's overall program, order leaders occasionally probed these activities for flaws and then offered constructive criticism to improve them. Pioneer efforts in this regard were carried on by the Ohio State Grange. An investigation by that body in 1881 revealed some alarming weaknesses in subordinate chapter social programs and prompted publication of a manual suggesting, among other things, ways to improve sociability at local grange sessions. First, it warned members to avoid making their "great feasts . . . overdone" because "every one [becomes] so wearied by preparations that they have no strength left for enjoyment." Second, it urged members to examine their rapport with their grange brothers and sisters to determine whether they were practicing the concepts of their brotherhood oaths sincerely or superficially. Finally, the handbook prodded grangers to treat each other in accordance with the Golden Rule.[44]

Occasionally, grange brotherhood pleas proved embarrassing because a few members took their "fraternal" vows too literally. On January 11, 1875, the Corona Grange of Mississippi expelled a matron and two male members for overindulging in brotherly affection. The woman allegedly was a "strumpet," and the men were reportedly her "paramores." A similar episode unfolded thirteen months later in Indiana at a session of Big Spring Grange No. 1963. On February 26, 1876, James Brim asked members of the chapter to eject Polly Ann Smith for deserting her husband and rendezvousing with Jonathan Turley at Bowling Green, Kentucky. The woman and her lover were both charter members of the same local, and presumably their intimate relationship developed as a direct result of their participation in granger activities. After a lengthy deliberation both individuals were

43 Alexandria (Va.) *Granger*, April 1, 1876; Chicago *Prairie Farmer*, April 22, 1876.
44 Madison *Western Farmer*, January 10, 1874; Ohio State Grange, *Hints and Helps*, 9–10.

found guilty of a "breach of good morals" and were ousted from the organization.[45]

On the positive side, granger brotherhood manifested itself in several ways. Many granges established committees of arbitration to resolve potential legal disputes between members. When this system operated according to plan, plaintiffs bypassed courts of law and took their litigations to chapter tribunals which presumably settled matters impartially without assessing any fees. Litigants not only saved attorney's charges, but they also were spared embarrassment of bringing public suits against their fraternal brothers.[46]

The best test of fraternal vows came during the 1870s when grasshoppers brought havoc to Great Plains farmers. By eating everything green in their paths, these insects left countless wheat belt families without any means of sustaining themselves or their livestock. Fortunately for the victims, however, assistance came from granges throughout the United States. Subordinate, state, and national chapters of the order poured thousands of dollars and tons of supplies into a region stretching from Kansas through Dakota Territory. In 1874 alone, the National Grange donated $3,000 to Kansas and $2,000 each to Nebraska and Minnesota. Undoubtedly, had it not been for these generous contributions from Patrons of Husbandry, locusts would have forced many more homesteaders to retire from this arid section than they did.[47]

As a rule, assistance came from grangers whenever they saw their brothers and sisters suffering. Patrons of Husbandry generously aided victims of the South Carolina earthquake of 1886, the Great Plains drought of the early 1890s, the central Georgia tornadoes of 1875, and the Mississippi River floods of 1874 and 1892. Within chapters themselves, relief committees comforted members who were ill and collected sundries, clothing, food, and money for families of deceased members.[48]

[45] Agnew Diary, December 14, 1874, January 11, 1875, Southern Historical Collection; Big Spring Grange, Secretary's Minutes, February–April, 1876.

[46] Chicago *Prairie Farmer*, August 23, 1873, June 16, 1877; Champaign Grange, Secretary's Minutes, February 8, 1874.

[47] National Grange, *Proceedings*, February, 1874, p. 37, 39; R. Q. Tenney to Saunders, June 6, 1875, in Saunders Collection; Iowa State Grange, *Proceedings*, 1873, p. 63; Nebraska State Grange, *Proceedings*, 1874, pp. 9, 11; Nebraska State Grange, *Grasshopper Circular* (Plattsmouth, Nebr., 1874); National Grange, *Proceedings*, February, 1875, p. 73; Ohio State Grange, *Proceedings*, 1876, p. 63.

[48] Lincoln *Nebraska Patron*, July 1, 1874; Raleigh (N.C.) *State Agricultural Journal*, June 4, 1874; Cottage Grange, Secretary's Minutes, June 11, 1874, Branchville Grange, Secretary's Minutes, May 6, 1875, in Branch Family Papers; Church Hill Grange, Secretary's Minutes, January 22, 1875, February 11, 1876.

Grange testimonials also tell of individual acts of charity. A typical display of brotherhood occurred in 1883, when a California granger and his wife were traveling by wagon in Indiana. Misfortune struck when the couple's team of horses jerked forward, tossing the woman into the air and out of the vehicle. Her fall onto the hard surface of the road broke three ribs, fractured her shoulder blade, and partially paralyzed her left arm. Fortunately, wrote the husband, her accident occurred near the home of grangers who, upon learning that she and her husband were members, invited the injured woman to stay with them until she was well enough to resume her westward trek. Speaking for his spouse and himself, the grateful husband stated in a public letter of appreciation published by the *Indiana Farmer* that "such kindness from those who so short a time ... [before] were entire strangers makes us feel that it is truly a good thing to be members of the Order of Patrons of Husbandry." [49]

From the outset, the more perceptive patrons had recognized that the true value of grange membership lay in the order's social and educational programs; their statements to this effect reflect what they expected to gain from grange membership. One order spokesman confessed that buying flour and bacon more cheaply, selling cotton more profitably, realizing economies in all legitimate ways, promoting cash transactions, and simplifying all business operations were "by no means unimportant objects ... [of grangerism], [but] we [grangers] shall make a great and in the end a fatal mistake, if we take these to be the sole or even principal ends of the organization. ... The Grange is primarily a social institution—a bond of union and a guarantee of good fellowship and kindly fraternal feeling. It brings together in its meetings the fathers, mothers, sons, and daughters of the neighborhood." [50]

In their eagerness to praise the Grange's record as a socializing agency, members often tended to exaggerate. Jonathan Periam, one-time farmer for the College of Agriculture of the University of Illinois and author of *The Groundswell: A History of Aims and Progress of the Farmer's Movement*—contended that "refining influences" of order membership had "prevented many a young man from spending his time and means in the village saloons, or billiard halls, and many soul-destroying resorts of vice." In addition, he said that the organization "has undoubtedly redeemed some who, but for its influence, would

49 Indianapolis *Indiana Farmer*, March 24, 1883.
50 Charleston *Rural Carolinian*, April, 1873.

have gone from bad to worse, and have died drunkards, and perhaps
have filled paupers' graves." [51]

Although Periam gave the Grange too much credit, his assessment
contained more truth about the societal value of grangerism and
about its meaning for those involved in the movement than is found
in the traditional historical works by Solon Buck and his disciples. To
patrons "the primary work of the Grange" was its "social and educational" pursuits.[52]

---

[51] Periam, *The Groundswell*, 147.
[52] Ohio State Grange, *Hints and Helps*, 5.

CHAPTER VII

Granger Business Attitudes
and Activities

The era following the Civil War witnessed an adjustment for American farmers. Changes precipitated in part by the bloody conflict's impact upon the United States' economy left the nation's rural residents staggering in a strange new world. The holocaust had brought high prices and prosperity, but these effects had vanished before 1870. Meanwhile, a new economic system based upon industry and the city was emerging to supersede agriculture and the farm as the predominant features of the American economy. This transformation bewildered rural inhabitants, and eventually their confusion engendered anger.

When Oliver H. Kelley and his associates were formulating plans for a national farmers' organization, they did not sense the new hostility and wrath arising in rural America. Their insulation from mainstreams of current agrarian thought resulted from their residence in the ivory-tower atmosphere of Washington, D.C. As a result, the original structure and objectives of the Grange lacked promise of deliverance from oppression and offered no concrete proposals for attacking agrarian economic problems.[1]

Thus, Kelley's first organizational effort was doomed before it began. As the Minnesotan soon realized, expectations of farmers rushing into the organization failed to materialize because in its original state the Grange was sterile in the eyes of those for whom it was intended. Farmers were interested primarily in "pecuniary benefit," and the Grange offered only intangibles. Kelley's subsequent conversations with prospective grangers firmly convinced him that success of the order was tied to billing the Grange as "an organization *for the protection of the farmer*" against monopolies and as a body committed to the economic improvement of its members. Consequently, the product of Kelley's direct contact with farmers was an organization with broader aims. The changes, however, were open-ended in the

---

[1] Kelley, *Origin and Progress of the Order*, 1–56.

sense that no exact details were formulated for effecting economic betterment. Definite programs came later and evolved gradually.

Grange experiments with various proposals for bringing about greater financial returns from farm products began early in 1869. The agent system and the cooperative mill were the first of a series of schemes developed by granges. They were suggested initially in a special report on order business possibilities presented at the 1869 session of the Minnesota State Grange. Preliminary plans called for operation of cooperative flour mills by subordinate chapters and for development of a network of state business agencies working in conjunction with a national business clearing office. According to the draft, the latter body was to serve as a marketing house and as a wholesaler of goods needed by farmers. Wheels of the grange business machine began moving at an informal meeting held after the first session of the Minnesota State Grange in 1869. A small group of state officers gathered at the home of C. A. Prescott and formed the first state grange business agency. To push the enterprise off the ground, the state hierarchy vested Prescott with the title of state agent and entrusted him with responsibility for obtaining items for members under the best possible terms. A request for a jackass was the first order handled by Prescott's office.[2]

No one was more favorably impressed with the initial actions taken by the Minnesota State Grange in behalf of cooperative business ventures than Kelley. In fact, the state chapter's announced plans made him very optimistic about the Grange's future. He now foresaw a day when every state grange might make arrangements with a variety of manufacturers and nurserymen, and Kelley also looked ahead to a time when state agencies might command enough strength to "break up the combinations between manufacturers and dealers" by bringing "the retail trade down upon" the Grange. In the founder's mind, such bright prospects would effect an overnight surge in order membership because crusades to eliminate middlemen "will wake up the farmers." On the other hand, in case the Grange's economic programs backfired, the national organization still had a safety valve. The parent body, reasoned Kelley, could always absolve itself altogether from the business activities of its subordinate chapters and deny an existence of any national sanctions approving these business pursuits.[3]

In the early 1870s when the First Granger Movement was at its zenith, the state agent system was being tested wherever the order

2 *Ibid.*, 112–13, 168, 180–81, 186, 242–75 *passim*, 302–303, 385–386.
3 *Ibid.*, 181, 186.

commanded sizable membership. Even though the National Grange never laid down specific guidelines, operational patterns of most grange business ventures were still basically the same. State organizations appointed one man to serve as state agent and gave him authority to contract with firms manufacturing items used by grangers and their families. Agents then were authorized to supply goods at the lowest possible cost to active members. For performing these services, agents received either a fixed salary or a commission based upon total monthly gross receipts. Regardless of payment system, holders of the position usually netted between $750 and $1,500 a year and received full expense accounts.[4]

In the initial stages, state agents secured discounts by a combination of salesmanship and perseverance. Firms handling goods needed by grangers were contacted and told of possible heavy volume sales through direct deals with Patrons of Husbandry. If these arrangements were sufficiently attractive, industrialists, nurserymen, wholesalers, and other suppliers informed state granges of their willingness to sell at prices below those available in retail markets. After gigantic trade possibilities had become clear, the process was reversed. Dealers then often sought out state agents and offered to sell products at special rates. Many concerns advertised their terms to grangers in order proceedings and in leading agricultural periodicals, and some even offered to sell to individual members directly. One such firm specializing in granger accounts was Montgomery Ward and Company of Chicago. It billed itself as the "Original Grange Supply House." Arrangements between state agents and businessmen enabled grangers to purchase a wide variety of goods at lower prices. Items commonly listed in agency catalogues and circulars included sewing machines and notions, musical instruments, fruit and shade trees, shrubs, seeds, fertilizers, agricultural implements and machinery, clothing, sundries, groceries, and myriad other useful items.[5]

How much money members actually saved by making purchases

---

[4] Iowa State Grange, *Proceedings*, 1873, p. 18; Indiana State Grange, *Proceedings*, 1874, p. 16; Wisconsin State Grange, *Proceedings*, 1875, p. 79; 1879, p. 71; 1880, pp. 61–65, 67; Michigan State Grange, *Proceedings*, 1876, p. 21; Illinois State Grange, *Proceedings*, 1875, pp. 16–7.

[5] Virginia State Grange, *Proceedings*, 1874, p. 35; Illinois State Grange, *Proceedings*, 1889, pp. 34–35. For advertising examples, see: Delaware State Grange, *Proceedings*, 1890, Appendix; Topeka *Kansas Farmer*, October 28, 1874. For a typical Ward's notice to grangers, see Lawrence *Spirit of Kansas*, August 10, 1876. Wisconsin State Grange, *Grange Agency Catalogue of 1877* (Milwaukee, 1877), 1–44; Indiana State Grange Purchasing Agency, *Circular* 7 (Indianapolis, n.d.), 1–4; Massachusetts State Grange, *Quarterly Bulletin*, January, 1876 (n.p.).

from state grange agencies instead of from local merchants is indeterminable because of a shortage of records. Moreover, the whole problem is compounded by the fact that circulars and catalogues distributed by state agents led members to believe that impossibly huge discounts resulted from dealing with grange cooperative enterprises. Dependent as they were for their positions upon annual votes of confidence given at state meetings, agents tended to exaggerate benefits accruing from these arrangements. Many agents boasted that they sold goods at prices which were as much as 50 percent below those available elsewhere. A typical report was made by an Ohio agent at the 1876 session of the state grange. According to his testimony, the Ohio state grange agency offered the following reductions to its customers:[6]

50%	off	on	sewing machines
25%	”	”	fruit trees
25%	”	”	footwear
30%	”	”	implements and machinery
20%	”	”	groceries
25%	”	”	lumber
5%	”	”	seeds
25%	”	”	sundries
45%	”	”	iron safes
45%	”	”	garden seeds
20%	”	”	dry goods

The bait of such huge savings evidently was effective because sales records of many agencies were very impressive. In an annual report given at the 1875 meeting of the state grange, state agent Alpheus Tyner of Indiana reported that business was brisk, and "orders for goods of every description have been pouring in from every section of the State, with a steadily increasing stream until the business has now assumed a magnitude scarcely hoped for by the most sanguine friends of the Order." Over $300,000 worth of goods had passed through his agency on the way to customers in Indiana and neighboring states during the previous fiscal year. Tyner's inventory showed that he had distributed:

agricultural implements
and miscellaneous

[6] James Cassidy to Osborn, March 19, 1873; Iowa State Grange Agency, *Confidential Circular*, 1873, Wisconsin State Grange Agency, *Strictly Confidential*, 1873, all in Osborn Papers; Indiana State Grange, *Proceedings*, 1875, p. 18; Ohio State Grange, *Proceedings*, 1876, pp. 33–34.

articles	.... valued at	$156,191.68
footwear	.... " "	41,000.00
groceries	.... " "	35,764.53
kerosene	.... " "	1,547.56
clothing	.... " "	8,710.28
stoves and tinware	.... " "	18,000.00
shelf hardware	.... " "	2,414.95
dry goods	.... " "	3,151.68
sewing machines	.... " "	43,000.00

Overall, Tyner estimated that collective buying had saved grangers 33 percent on their purchases. Moreover, he reported that the past year's operating expenses of the agency had only amounted to $6,087.15. Of this figure, Tyner and his clerk received $3,183.33 in salaries, rent consumed $1,599.95, and incidental expenditures totaled $1,303.87.[7]

Figures for other agencies were equally noteworthy. Business had boomed so much in excess of expectations in Ohio that cooperative enterprise there experienced growing pains. When agency sales were soaring toward the $1 million mark, the agency found that it had outgrown its facilities at Sharonville. This called for transfer, so the state grange moved its business headquarters to a large warehouse in downtown Cincinnati. Even in states where the order was not especially strong, cooperation was tried on a smaller scale but with seemingly good results. For example, volumes of trade negotiated by the Oregon and Michigan state agencies exceeded $45,000 a year in the mid-1870s, and neither of these states were considered centers of grangerism. Taken altogether, grange cooperative enterprises represented a total estimated business volume of $18,000,000 in the spring of 1876.[8]

Contrary to impressions left by Solon Buck in his study,[9] state agencies did not always become insolvent with the collapse of the First Granger Movement and they did not necessarily contribute directly to the order's decline in the late 1870s. Business records of the Wisconsin State Grange offer the best refutation of Buck's theses. Agency statistics for the north-central state leave no doubt that the principle of cooperation continued to produce savings in the face of a steadily dropping membership. In 1875—the peak year for membership—agen-

7 Indiana State Grange, *Proceedings*, 1875, p. 18.

8 Ohio State Grange, *Proceedings*, 1876, p. 29; Oregon State Grange, *Proceedings*, 1875, p. 9; 1876, pp. 45–46; Michigan State Grange, *Proceedings*, 1877, p. 32; Chicago *Industrial Age*, May 6, 1876.

9 Buck, *The Granger Movement*, 260, 262, 274–77.

cy sales were at a low point, but later there was an inverse relationship between sales and membership.

YEAR	MEMBERS	SALES	AVERAGE AMOUNT SOLD PER GRANGER
1875	18,653	$ 38,194.39	$ 2.04
1876	18,427	115,882.31	6.29
1877	17,640	164,445.16	9.32
1878	7,093	86,391.92	12.18
1879	5,526	61,334.44	11.08
1880	4,651	55,560.20	11.93
1881	3,960	NO DATA AVAILABLE	(NDA)
1882	3,385	56,992.60	16.84
1883	3,217	67,248.23	20.90
1884	2,934	63,472.97	21.63
1885	2,439	56,976.53	23.36
1886	2,343	48,573.76	20.73
1887	NDA	34,023.97	NDA
1888	*	54,815.84	NDA

* Membership increased by eight over the 1887 figure.[10]

State agencies not only sold assorted goods to grangers, but they also marketed members' farm products with an expectation of obtaining higher prices. In the South, grange cooperatives attempted to place staple growers in a better bargaining position by receiving their cotton and holding it until the market price improved. This scheme was largely unsuccessful because it was distasteful to the American agrarian tradition. There were always enough growers who placed the success of the scheme in jeopardy by refusing to pool their marketing efforts. For example, when state agencies of Louisiana and South Carolina respectively had in 1875 and 1879 only 4,797 and 1,710 bales at their disposal with which to produce an upward swing of cotton prices in those two years, no one could expect miracles from the two grange cooperatives, and there were none.[11]

Outside the Cotton Belt, experiments with cooperative marketing were tried with other products. Grangers from West Virginia, Ohio, and Pennsylvania joined forces in the 1870s and formed a tristate wool growers association with headquarters at Steubenville. The ex-

[10] Wisconsin State Grange, *Proceedings*, 1881, pp. 37–40; December, 1882, p. 52; 1883, p. 28; 1884, p. 19; 1885, pp. 18–19; 1886, p. 18; 1888, p. 14.
[11] South Carolina State Grange, *Proceedings*, 1879, n.p.; Louisiana State Grange, *Proceedings*, 1875, pp. 28–30.

change sold sheep, marketed raw wool, and distributed woolen cloth to patrons. The association operated until the end of the decade and then died from lack of patronage.[12] Meanwhile, grange agencies elsewhere marketed farm products with varying degrees of success during the last three decades of the nineteenth century. The Iowa State Grange employed a resident agent at the Union Stock Yards in Chicago during the early 1870s, and his sales record was exemplary. In 1873 his receipts exceeded $200,000. Excellent as it was, the Iowan's performance was still below accomplishments registered eighteen years later by the Maryland State Grange agency. It received and moved $260,485.19 worth of farm products, including 70,927 bushels of grain, 1,351 hogsheads of tobacco, 12,789 bushels of potatoes, and 10,139 pounds of wool.[13]

Grangers tried marketing their own grains, fibers, and animals largely because they felt warehouse, elevator, and stockyard proprietors conspired against farmers to depress prices paid producers. At the same time, they hoped their cooperative efforts would destroy the individuals responsible for "futures." Members of the order charged the "gamblers" who controlled boards of trade with upsetting "the old law of supply and demand regulating prices" and with making the economic principle a "myth." Their accusations were based on a belief that biddings on unfattened livestock and unharvested crops predetermined prices of meats, grains, and cotton. Attempts at collective marketing soon taught grangers that their cooperative programs were not going to rescue farmers from abusive practices of dealers in futures, so the order shifted its fight to Congress. Patrons now pushed for solons to "pass a law making it a felony for anyone selling their promises to deliver goods that they have not and never expect to have." [14]

In the early days, grange business ventures faced many obstacles,

12 J. D. Whitham to Clarke, May 25, November 24, 1877, January 29, September 23, 26, December 6, 1878, March 22, October 22, 1879, Receipt from Wool Growers' Exchange, January 23, 1879, all in Clarke Papers.

13 Oregon State Grange, *Proceedings*, 1875, pp. 7, 36–37; Ohio State Grange, *Proceedings*, 1876, pp. 29–30; Illinois State Grange, *Proceedings*, 1891, pp. 64–65; Champaign County (Ill.) Grange, Secretary's Minutes, December 21, 1875, January 11, 1876, Champaign County Grange Record Books, Illinois Historical Survey Collections, University of Illinois Library, Urbana; Virginia State Grange Business Bureau, *Confidential Circular*, July, 1877 (Lynchburg, 1877); Chicago *Industrial Age*, December 20, 1873; Maryland State Grange, *Proceedings*, 1891, p. 10.

14 National Grange, *Proceedings*, 1886, p. 75; Wisconsin State Grange, *Proceedings*, 1890, p. 8.

and some of these resulted from exaggeration of benefits. Organizers went considerably beyond practical limits to draw additional farmers into the organization and to bolster state agency sales. Although immediate consequences of false claims were seemingly harmless, their long-range ramifications were not. Members who had expected miracles from cooperation when they joined the order soon discovered from their experiences that they had been deceived. High expectations thus converted confidence into mistrust and disenchantment, and these feelings of disillusionment eventually led to mass withdrawals from the Grange.[15] Overextension of cooperative activity was another noxious effect resulting from false claims. Convinced that they had found a way to cure their financial ills, grangers became receptive to almost every scheme calling for a pooling of efforts. As a result, patrons outdid themselves by approving myriad plans which were impractical and ill-conceived. At the same time, state grange conventions virtually turned into planning orgies where delegates hastily concocted panaceas for ridding farmers of their dependence upon retail merchants. In fact, almost any proposal prefaced with remarks denouncing middlemen as villainous vampires was certain to receive serious consideration if not immediate approval—regardless of the scheme's lack of merit or feasibility.

Grange receptivity to poor plans and to harangues ringing with class consciousness was never more obvious than it was at the 1875 session of the Montana Territorial Grange. In explanation it is not difficult for us to see where agricultural marketing in Montana Territory was unique in many respects. The economy was that of a frontier area, and inadequate transportation development caused distribution problems. As a result, all markets of consequence were found within the territory and they were created principally by mines, stage lines, and government contracts and were controlled by a band of price-dictating middlemen. It was this latter fact which most annoyed Montana grangers. Their chief complaint was that Montana middlemen were parasites living off the labors of the territory's miners, mechanics, and yeomen while offering nothing in return:

> Our middle men live in fine houses, expensively furnished, their tables are spread with the best luxuries of our country and luxuries imported from abroad. Their sons and daughters are raised in ease and affluence, become accomplished and graduate at colleges and seminaries. The dru[d]gery of the middle man's kitchen is performed by a servant,

[15] Wisconsin State Grange, *Proceedings*, 1880, pp. 61–65.

while his wife and daughter, dressed in silks, with jeweled rings upon their delicate fingers, preside in the parlar [sic]. ... But how, or by whom, is the expense of this "high stile" paid? By the labor of the farmer and the miner, the principal producers of our Territory, taken from as the profits of so-called legal trade.

In sharp contrast, "producers" were pictured as a class living in simplicity and deprivation. Their homes were log cabins, "and their wives preside[d] in the kitchen, sitting-room and parlor all in one, clad in callico [sic], and with the rough impress of labor upon their hands." At the same time, children of miners and farmers only obtained rudiments of an education at one-room schoolhouses.[16]

Class lines having been drawn, Montanans were ready for the next step in a plan's presentation, and it was familiar. After sufficient emotional conditioning, grange audiences would be offered programs which promised an end to oppression and despair. The remedy presented to Montana grangers in 1875 had the same ring of deliverance as other proposals suggested elsewhere, but the Montana program for making area farmers more independent and more financially solvent had more Marxian overtones than usual. It called for the territory's proletariat—the miners and farmers—to stand together and revolt collectively against the vile system of middlemen. The plan revolved around the creation of a federation of producers whose powers included receiving and marketing farmers' produce.[17]

On the surface, the Montana plan offered territorial farmers relief from their financial headaches. However, hidden beneath the panacea's veneer was a simple sugar pill whose total curing punch proved negligible when faced with Montana's complex economic ills. Weakness became transparent with attempts to make the weakly conceived plan operative. Montana grange officials, like others before and after them, found that their plan's success had been too tightly tied to cooperation—a rare quality among American farmers. Also lacking in the plan was consideration of how the general nature of participants could be changed. Taken together, this proposal was nothing but a product of haste, emotionalism, oversimplification, and ignorance.[18]

In these respects, however, the Montana plan was not really dissimilar to many other schemes for cooperation tried elsewhere in the 1870s. Most were the handiwork of individuals unfamiliar with rural

16 Montana Territorial Grange, *Proceedings*, 1875, pp. 17–19.
17 *Ibid.*, 18–19.
18 *Ibid.*; Massachusetts State Grange, *Proceedings*, 1899, p. 89.

habits, basic business principles, and elementary entrepreneurial practices. These unenlightened and inexperienced souls correctly saw certain elements in society allied against them, so they polarized people into two rival factions—the oppressed and the oppressors. The prevalence of these opinions and factors explained in large part why so many absurd proposals won granger approval in the late nineteenth century. They offered a place for the downtrodden at the top, and they were couched in emotional terms. The result of this mass psychopathy was a wave of plans which were ill-conceived and impractical. Not enough time went into study of socialist theory for the plans to have sound groundwork and for members to understand the struggles ahead.[19]

Of course, business ventures operated by state granges had many other visible weaknesses. Notably, state agents had no means at their disposal to gauge future buying habits of their customers. As a result, agencies often overstocked merchandise, or they kept inventories below levels needed to fill orders promptly. Gradually, however, the supply problem corrected itself. Granges learned to pool their orders and to forward to agents annual estimates of their members' needs. Earlier agency sales also helped grange cooperative managers to gain the experience necessary to operate their businesses more efficiently. The inexperience of grangers as businessmen showed itself best in 1874. In that year, a Chicago man posed as a sewing machine manufacturer and made the rounds of Illinois granges taking orders. Months passed before his victims realized that they had been swindled of an undisclosed amount of money.[20]

The net supply of circulating currency per capita declined sharply in the late nineteenth century. Tight money naturally had a bearing on consumers' purchasing abilities. It also became increasingly difficult to provide credit arrangements. Specifically, the currency shortage made it necessary for most agency transactions to be on a cash-only basis. There was not enough working capital available in agency cof-

[19] Chicago *Western Rural*, May 29, 1875; Michigan State Grange, *Proceedings*, January, 1875, p. 17; George Cerny, "Cooperation in the Midwest in the Granger Era, 1869–1875," *Agricultural History*, XXXVII (October, 1963), 190.

[20] Illinois State Grange, *Proceedings*, 1889, p. 46; 1892, p. 33; Wilmington Grange, Secretary's Minutes, January 15, 1876; Whitham to Clarke, April 28, 1879, in Clarke Papers; Alex J. Wedderburn to H. L. Couchman, April 1, 1887, in George Couchman Family Papers, West Virginia University Library, Morgantown; Dragoon Grange (Osage County, Kans.) No. 331, Secretary's Minutes, February 27, 1874, Kansas State Historical Society Library, Topeka; Charleston *Rural Carolinian*, July, 1874.

fers to permit credit sales without dipping deeply into reserves used for obtaining commodities and manufactures from wholesalers who demanded cash. Even had surplus funds been available, it is doubtful whether agencies would have allowed their customers to buy on credit. Grange leaders generally were dismayed by the concept of not paying for purchases at the time of delivery. To them, it was an ugly form of usury which held debtors in bondage to creditors.[21]

Many problems affecting grange businesses were directly related to the enterprises' varying financial structures. Although no two ventures were managed exactly alike, trends nevertheless existed. For purposes of simplification, grange cooperatives in the late nineteenth century can be grouped into two basic categories. Of the two, the one gaining widest acceptance was the organizational plan which made business operations an arm of the state grange. Revenues from general treasuries got these businesses off the ground and supplemented marginal profits; state granges sat as governing boards. In contrast, some state chapters chose to sever their business activities from the general organization. They created independent establishments whose main duties were selling and marketing. Operating upon England's Rochdale plan, these firms raised capital from private sales of stock to grangers, and direction and management came from boards of trustees elected periodically by shareholders. Care was usually taken to keep these enterprises free from monopolistic tendencies. To this end, limitations were often placed on the number of shares of stock sold to individual buyers.[22]

At first, the National Grange showed no organizational preference, but signs pointing to disenchantment with agent systems became evident at the 1875 session. The parent body suggested a strengthening program which, had it been adopted, might have saved the agent system. In essence, it provided for more supervision from the top, greater concentration, and less duplication. The plan, however, was never pushed. Rather than connect itself with failing state agencies, the National Grange disavowed its support for the agent system in 1877 and

---

21 Nebraska State Grange, *Proceedings*, 1874, p. 23; Herbert S. Schell, "The Grange and the Credit Problem in Dakota Territory," *Agricultural History*, X (April, 1936), 71; Mildred Throne, "The Grange in Iowa, 1868–1875," *Iowa Journal of History*, XLVII (October, 1949), 308–309; St. Louis *Colman's Rural World*, March 21, 1874.

22 Denver *Rocky Mountain Weekly News*, March 4, 1874; New Hampshire State Grange, *Proceedings*, 1877, p. 17; Wisconsin State Grange, *Proceedings*, 1875, p. 60; Throne, "The Grange in Iowa," 308.

began a campaign among its subordinates for establishment of cooperative stores patterned after Rochdale enterprises of Great Britain.[23]

Some chapters were already in the process of setting up cooperative stores when the National Grange directive reached them, while others had been running such businesses for years. The most ambitious project was one in the lower Mississippi River Valley. The state granges of Arkansas, Tennessee, Louisiana, and Mississippi combined to form the Southwestern Co-operative Association, whose purposes included establishing and running cooperative stores in the four-state region. Sales of stock to grangers financed the association, with precautions to involve as many members as possible in the project. A charter rule specified that stockholders were not to hold more than two hundred shares in the association, and capitalization was fixed at $50,000.[24]

The Texas Co-operative Association was one of the best-planned cooperative enterprises established by Patrons of Husbandry. Its record of achievement was exemplary. The association remained active for almost two decades, survived two depressions, and paid regular dividends on its $100,000 capitalization. Its earning rate was especially good in the 1880s. During the fiscal year 1882–1883, the association divided among its stockholders profits totaling $22,023.81.[25]

Although grange proceedings seldom alluded to reasons for agency failures, the financial structures of many cooperatives do suggest a few weaknesses. Ventures dependent upon stockholders' confidence and steadfastness more than likely collapsed as a result of rapid withdrawals of capital resulting from hard times and disenchantment. Economic depressions had less impact upon those agencies connected with state granges. The latter arrangement's success was tied more directly to agency efficiency and to state grange viability.[26]

Lack of direction and coordination were two other problems weakening grange cooperative efforts. Each state grange acted alone, and overlapping resulted. Attempts at unison were tried, but they usually bogged down because state grange officers refused to surrender their

[23] National Grange, *Proceedings*, February, 1875, p. 26; 1877, p. 17. Agent Joseph H. Osborn of Wisconsin saw the need for the same basic changes late in 1873, but no one heeded his advice. Osborn to John Samuel, December 18, 1873, in John Samuel Papers, Wisconsin State Historical Society Library, Madison.

[24] Chicago *Industrial Age*, March 4, 1876; Topeka *Kansas Farmer*, January 28, 1874; Chicago *Western Rural*, December 23, 1876; Douglas County Pomona Grange, Secretary's Minutes, March 25, 1876; Charleston *Rural Carolinian*, October, 1876.

[25] Texas Co-operative Association, Patrons of Husbandry, *Proceedings*, 1880–1897.

[26] National Grange, *Proceedings*, February, 1875, p. 26; *ibid.*, 1877, p. 17; Illinois State Grange, *Proceedings*, 1875, p. 15.

authorities and the National Grange provided too little assistance. Even when conventions met, little that was substantial was ever accomplished. Typical were the meetings held in 1875 and 1894 by midwestern grange business officials to iron out difficulties for setting up a regional body. True to form, neither session produced anything except many meaningless resolutions pledging greater unity. By not doing more for interstate cooperation, grange leaders spoiled their chances for promoting economic democracy.

Splits over policies also weakened the "cooperative" work of granges. Some agents felt that their hands would be strengthened if they were empowered to sell goods to all farmers, but dogmatic members adamantly refused to permit extensions of agency services to nonmembers. They clung to the belief that the right to use agencies belonged only to members of the order because outsiders contributed nothing to the movement. The outgrowth of these feelings was battles between grangers who wanted insulated agencies and ambitious agents who sought sales to other potential buyers. Two notable confrontations over this matter divided Minnesota and Wisconsin grangers in 1873 and 1880 respectively. Agent J. S. Denman of Minnesota inadvertently touched off the first controversy by proposing that all farmers be permitted to buy through the state agency. Denman's suggestion did not become a dead letter; instead, it triggered an immediate reaction which set him against members who disagreed with his position. In the dispute that followed, neither side respected the other's opinions; damaging ramifications resulted from which the order in Minnesota never recovered. Denman's critics accused him of betraying his grange brothers and sisters. They also demanded his removal. Had Denman been wise, he would have ignored his opponents' charges. Unfortunately, Denman's determination to have the last word and his hostility to criticism were greater than his propensities for reconciliation. As a result, Denman plunged himself into deeper trouble with his emotional outbursts. In an open letter to readers of the *Farmers' Union*, Denman scoffed at his opponents and challenged their sanity. In the same letter, he sarcastically called for an agreement between the state mental asylum and the state grange because it was the "duty" of the order "to take care of . . . its own Lunatics." [27]

[27] Alpheus Tyner to Osborn, July 29, 1875, L. G. Kniffen to Osborn, August 14, 1875, both in Osborn Papers; Illinois State Grange, *Proceedings*, 1894, pp. 10–11; Minneapolis *Farmers' Union*, November 1, 8, 15, 1873; Wisconsin State Grange, *Proceedings*, 1880, pp. 42, 61–65.

A similar situation developed seven years later in Wisconsin. At the 1880 session of the state grange, the Dodge County Pomona Grange sharply attacked L. G. Kniffen's handling of the state agency. The agent was criticized because he had opened agency services to non-members. Initial charges then set off a chain reaction of resolutions demanding Kniffen's ouster and an investigation of grange business operations. The check that followed disclosed that agency goods cost 10 to 33 percent more than comparable merchandise at leading Milwaukee firms. Despite anger produced by Kniffen's policies and evidence showing his inefficiency, no specific actions were taken against him. Nevertheless, the whole affair drove another wedge into the order's efforts to produce fraternalism among farmers.[28]

On a few occasions, dishonesty also affected grange cooperatives. Of the crimes committed, embezzlement was the most serious. The largest reported loss occurred in 1874 when the treasurer of the Missouri State Grange speculated with order funds and lost $20,000. The gambling spree did not bankrupt the chapter, but it forced drastic reductions in cooperative programs. The Wisconsin State Grange was not as fortunate. On May 1, 1890, L. G. Upson sued the state chapter for failing to pay him for goods delivered to the state agency managed by L. G. Kniffen. Although the lawsuit was finally settled by demurrer, it still ruined the state agency.[29]

Grange cooperatives constantly faced the problem of selling inferior merchandise. Established industries like the McCormick Reaper Company did not necessarily have to open new accounts to remain solvent, whereas many new ones did. Consequently many products handled by agencies were made by fly-by-night newcomers to industry, who had not built up reputations for excellence. Most agents hoped for the best, and indiscriminately sold goods to their unsuspecting brothers without knowing anything of the item's worth. If complaints were sufficiently numerous, a line of products was discontinued. The consequences of these methods were many dissatisfied buyers and much lost prestige for the order.[30]

When businessmen made concessions to grange cooperatives, they expected members to keep terms secret. Nevertheless, many patrons succumbed to temptation and divulged conditions of confidential con-

28 Wisconsin State Grange, *Proceedings*, 1880, pp. 42, 61–65.
29 Madison *Western Farmer*, November 28, 1874; Wisconsin State Grange, *Proceedings*, December, 1890, pp. 16–17; 1891, p. 32.
30 Indiana State Grange, *Proceedings*, 1874, p. 37; South Carolina State Grange, *Proceedings*, 1875, n.p.

tracts. Their failure to guard price information placed manufacturers in obviously awkward situations, causing many firms to terminate relations with granges. Concerns that had cautiously watched developments before acting now decided not to negotiate with grangers because the drawbacks of deals with them seemingly outweighed the benefits.[31]

Inducing manufacturers to sell their goods at discount prices to grangers was a big challenge to all types of agencies. It was understandable that many industrialists ignored patrons' requests for special rates. There were no assurances that agreements with cooperatives would result in benefits, but there was much to suggest that such arrangements would have damaging ramifications.

Wise entrepreneurs knew how valuable the willingness of middlemen to grant special payment terms was to sales. In the heavy farm equipment alone, 60 percent of the buyers in the 1870s needed credit to finance their purchases. At the same time, business leaders were aware that retail dealers usually were prepared to service what they sold, and they also recognized that their franchises kept prices at a steady level above the cost line. On the other hand, the only possible returns offered manufacturers from deals with grange agents were ones with purely speculative overtones. Thus, with no guarantees and with possible losses staring them in the face, many prudent industrial leaders spurned agreements with the farmers' order.[32]

Manufacturers' rebuffs often produced angry reverberations at grange sessions. Chapters rebuked unrelenting companies and threatened them with boycotts. When firms laughed at grange threats of reprisal, state and local chapters responded with a multitude of resolutions urging grangers to refrain from buying anything manufactured by blacklisted businesses. At the same time, names of specific enterprises guilty of repulsing grangers were published and circulated to grange "organs" and to patrons.[33]

Grange retaliatory measures did not always stop, however, with boycotts; the order occasionally struck back by organizing its own

[31] Michigan State Grange, *Proceedings*, January, 1875, pp. 17–18.

[32] Throne, "The Grange in Iowa," 305–307; A. H. Hirsch, "Efforts of the Grange in the Middle West to Control the Price of Machinery," *Mississippi Valley Historical Review*, XV (March, 1929), 476–96; Illinois State Grange, *Proceedings*, 1891, p. 49; St. Louis *Colman's Rural World*, March 21, 1874.

[33] Alexandria (Va.) *Granger*, April 1, 1876; Iowa State Grange, *Proceedings*, 1873, p. 51; Chicago *Industrial Age*, February 14, 1874; Wisconsin State Grange, *Proceedings*, January, 1874, pp. 8–9; Columbus (Miss.) *Patron of Husbandry*, August 23, 1879.

manufacturing concerns to compete with existing firms. Initially, farmers' industrial cooperatives grew out of abortive attempts to obtain harvesters at discount rates. In 1873 the Iowa State Grange literally exhausted itself trying to wring price concessions from harvester producers, but not one yielded. As a result, the frustrations of 1873 drove the state grange executive committee into the manufacturing business. To this end, the order purchased Amos G. Warner's harvester patent for $1,300 and established assembly plants at Boone, Des Moines, Osage, and Oskaloosa.[34]

These grange harvester works in Iowa had an erratic career. At first, it looked as if cooperative manufacturing was destined to succeed. Grange-produced machines were selling for 50 percent less than competitive implements, and grangers were swamping their factories with requests. Testimony to the early success of the operation was the backlog of over a thousand orders on March 15, 1874. The good fortunes of spring disappeared, however, almost as soon as the harvesters appeared in the fields of their owners; word spread rapidly across the state that grange machines were defective. A bad reputation dropped sales so rapidly in 1875 that the cooperative business lost $1,008.11 and bankrupted the state grange.[35]

This manufacturing attempt in Iowa was only one of a number of industrial projects put into operation by granges in the seventies. During the decade, cooperative manufacturing was especially evident in Missouri. On May 12, 1874, the voice of the state grange boasted that Missouri grangers had already planned or completed eighteen grain elevators and warehouses, eight gristmills, three meatpacking plants, and twenty-six manufacturing establishments. The latter concerns produced everything from plows to cheese.[36]

Not even the conservative Deep South was immune to suggestions for industrial cooperatives. In fact, some of the most radical proposals originated in Dixie, and all were offered by grangers long before the elections of Tom Watson and Pitchfork Ben Tillman, symbols of the "revoltin' rednecks." At the 1875 session of the Georgia State Grange, a special committee on manufactures advocated a complete overthrow of existing economic structures. Using statistics to support its position,

---

[34] Cerny, "Cooperation in the Midwest," 200, 202; Charleston *Rural Carolinian,* October, 1874; Throne, "The Grange in Iowa," 309–10; Wisconsin State Grange, *Proceedings,* January, 1874, p. 21.

[35] Cerny, "Cooperation in the Midwest," 202.

[36] Macon *Missouri Granger,* May 12, 1874.

the committee showed that it was feasible to establish granger-owned cotton mills in every county in the state and that a minimum capital investment of $22,500 was adequate to operate a profitable textile factory with twelve hundred spindles.

DAILY DEBITS

1,000 pounds of cotton at 11 cents	$ 110.00
Labor and supplies for spinning	40.00
Commission for selling	13.50
Depreciation	3.00
Insurance on $15,000 at 3 per cent	1.50
Total expenses per day	$ 168.00

DAILY ASSETS

Daily production of 900 pounds of No. 10 yarn selling at 20 cents	$ 180.00
Total earnings per day	12.00
Total earnings per year	$3,600.00
Annual profit return	16.5% [37]

The object of this plan was granger control of the fiber-to-cloth process. Although Georgia grangers never invested their money in cooperative cotton mills on a scale as large as recommended by the committee on manufactures, their willingness to explore new possibilities for worker-owned industries still suggested that they, like other farmers in the United States, opposed unbridled capitalism and welcomed any plan offering relief.[38]

A survey of grange-operated industries discloses that most cooperative manufactories run by order chapters were dependent exclusively upon the organization's own resources. Revenues obtained from closed sales of stock and grange treasuries launched the operations, while small inner cliques of directors and officers provided leadership and control. Aside from the managerial and financial sides of these concerns, granges deliberately shunned outsiders in the process of operating businesses. It seems that self-willed individuals who put these industries into effect were convinced by their own rhetoric that running plants profitably required no particular skill or ingenuity. Given the wide prevalence of such feelings, it was understandable that

[37] Georgia State Grange, *Proceedings*, 1875, pp. 24–25.
[38] *Ibid.*, 24–29.

grange leaders avoided involvement with nonfarmers thought to be lacking in sympathy for agrarian objectives.[39]

There were a few departures from the normal pattern. In the Southwest, plans were set September 7, 1875, for the organization of the Texas Grange Manufacturing Association, an unusual cooperative based somewhat on Rochdale principles. The association was conventional in the sense that sales of $25 shares of stock to members capitalized the project at $250,000, but it was different in that grangers did not have controlling interest in the concern. Industrialist George A. Kelly was a majority-right partner. In exchange for an agreement calling for conversion of his foundry into a factory, Kelly received special dispensations which gave him primary control. After four years of joint ownership, Kelly dissolved the arrangement by buying out his partners.[40]

Another example of grange industrial cooperative ingenuity existed at Clarksville, Tennessee. In September, 1874, Publisher M. V. Ingram of the *Tobacco Leaf* spoke to grangers at Noah's Spring Exhibition, telling them of a scheme by which members of the order hopefully would be able to purchase farm vehicles at considerable savings. The inventor of Ball's Wrought Iron Wagon lacked money to build a wagon works, so Ingram proposed an arrangement by which the developer would have capital and the city of Clarksville would have a new industry. All grangers had to do, noted the newspaperman, was raise the funds, and the plant would locate in the northern part of middle Tennessee. Even more lucrative for farmers were the prospects of placing the inventor at the mercy of grangers by giving him financial assistance. Ingram did not have to remind grangers that holding purse strings would place members of the order in a good position to demand wagons at a fraction of the retail price. The final product of these efforts was a tentative three-way partnership between the grangers who bought stock, the community, and the inventor.[41]

Whenever grange industrial cooperatives failed, it was common practice for order spokesmen to evade true causes of failure. Along these lines, grangers claimed that new legislation was needed to curb

[39] Indiana State Grange, *Proceedings*, 1891, p. 6; 1898, pp. 9–10; 1899, p. 11; New York State Grange, *Proceedings*, 1891, pp. 124–25; Pennsylvania State Grange, *Proceedings*, 1887, pp. 11, 16–21; Illinois State Grange, *Proceedings*, 1891, pp. 70–71; New Hampshire State Grange, *Proceedings*, 1874, pp. 35–36.

[40] Ralph A. Smith, "The Co-Operative Movement in Texas, 1870–1900," *Southwestern Historical Quarterly*, XLIV (July, 1940), 34–36.

[41] Clarksville (Tenn.) *Tobacco Leaf*, September 9, 16, 1874.

giant trusts that strangled competition, including that provided by grange manufactories. The laws proposed in the 1870s were aimed at existing patent rights. Members of the order claimed that privileges afforded holders of patents stifled competition and created natural monopolies. The grangers' remedy called for two basic changes. First, patent rights should always remain open to anyone who wanted to purchase them. Second, patentees should receive a fixed royalty from every company, cooperative, or individual who borrowed a new method, design, or product.[42]

What made this proposal unique was not so much the plan itself, but the shift in philosophy it represented. There was a growing contention among grangers that the federal government had to be active because it had a responsibility to insure justice and equal opportunities for all citizens. In effect, leaders of the order were calling for an extension of the arm of government into new spheres of activity.

Subordinate and pomona chapters frequently augmented cooperative activities of state granges. Literally hundreds of local and county branches either established Rochdale cooperative stores and agencies or employed special business agents; these purchasers and mutuals usually followed the same basic procedures and had strengths and weaknesses similar to their state-level counterparts. If agents were hired, they solicited discounts from area businessmen and pooled orders to acquire rate reductions and to reduce shipping costs. Several local cooperatives were mutual stock companies. Using the Rochdale plan as their models, these local cooperatives adopted constitutions whose bylaws related to such matters as capitalization, officers, election procedures, dividends, and stock limitations. Survival of these mutual businesses was largely dependent upon sales of stock to members. In most cases, early enthusiasm generated by the cooperatives' prospects was enough to promote sufficient sales, but there were instances when grangers refused to invest adequately in their enterprises. In Tolono, Illinois, eight grangers each lost approximately $1,000 because they were the only subscribers for a mutual warehouse project.[43]

---

[42] Wisconsin State Grange, *Proceedings*, 1875, pp. 60–61; Iowa State Grange, *Proceedings*, 1873, p. 45; Michigan State Grange, *Proceedings*, January, 1875, p. 42; J. A. Swain to Osborn, May 15, 1875, J. D. Whittet to Osborn, July 1, 1875, both in Osborn Papers.

[43] Chicago *Prairie Farmer*, May 22, 1875; Macon *Missouri Granger*, March 23, 1875; New York State Grange, *Proceedings*, 1892, p. 55; T. M. Brantley to Lawrence J. Guilmartin, August 25, 1875, in Lawrence J. Guilmartin and John Flannery

Besides operating agencies and Rochdale-patterned stores handling merchandise, local and pomona granges also developed storage facilities and market services. In theory, mutually owned grain, tobacco, and cotton warehouses were to assist grangers in two ways. First, they were to store members' products at low rates. Second, they were to combine sales of goods in order to enhance the growers' bargaining position in the market. The latter principle was also to apply to collective sales of livestock, as well as to field products. In practice, however, grangers never accomplished their objectives because they could not control enough of the total market to effect an upward price trend. This was true even in Iowa, a state in which one-third of the grain elevators and warehouses were once owned by order chapters and from which five million bushels of grain were shipped to Chicago in 1873 on one account.[44]

Although the same rules were practiced, many local businesses operated more efficiently than their state counterparts. While several state granges were caught up in reams of red tape, many local and county agencies had streamlined operations which minimized overhead expenses and kept inventories at a minimum. This, combined with regular adherence to a practice of only ordering those goods which had been requested, enabled some subordinate and pomona agencies to run smoother than state businesses and helped them to pass on consistently greater savings to members.[45]

With few exceptions smaller cooperatives had some advantages, but generally the same types of problems plagued subordinate and

Papers, Duke University Library, Durham; Raleigh Grange No. 17 Letters, Duke University Library Collection, Patrons of Husbandry; Champaign County Pomona Grange, Secretary's Minutes, November 3, 10, 21, 1873; Champaign County Pomona Grange, Farmers Coopperation [sic] Association, Record Books, April 13, 1875; Dunn County Wis.) Pomona Grange, Constitution and By-Laws of the Co-operative Council of Dunn County Patrons of Husbandry (Neenah, 1875), 1; Elm Creek (Tex.) Cooperative Association, Charter and By-Laws (Sequin, 1891), 6–13; Madison Western Farmer, December 20, 1873.
44 Illinois State Grange, Proceedings, 1896, pp. 59–60; Olive Grange, Secretary's Minutes, November 13, 1886; Church Hill Grange, Secretary's Minutes, December 27, 1878; Chicago Prairie Farmer, January 8, 1876; Topeka Kansas Farmer, January 24, 1877; Virginia Senate Journal (1875), 310–11, 337, 343, 385, 446; Brownville (Nebr.) Nemaha County Granger, September 15, 1876; Charleston Rural Carolinian, June, 1873.
45 A sampling of savings will be found in the following sources: Massachusetts State Grange, Proceedings, 1898, p. 108; Indianapolis Indiana Farmer, February 12, 1881; Charleston Rural Carolinian, October, 1872; Church Hill Grange, Secretary's Minutes, January 25, 1878.

Discing the fields in Barton County, Kansas, in the 1890s

Horse-powered hay stacker in Kansas in the 1890s

Heading wheat in Kiowa County, Kansas, in the 1890s

Meeting of Grangers, from *Frank Leslie's Illustrated Newspaper,* August 30, 1873

County Fair in Hennepin County, Minnesota, 1890

County Fair in Hennepin County, Minnesota, 1890

The Patrons of Husbandry letterhead stationery showing a grange meeting in progress

The Grange Room with officers in position, from *The Grange Illustrated; or Patron's Handbook* by John G. Wells, New York, 1874

Exterior view of Oliver Kelley's house

Illustration of Grange officers found in *Frank's Leslie's Illustrated News-paper*, September 13, 1878.

Mrs. Mary Bryant Mayo, zealous leader of the women's movement

Putnam Darden

Darden Monument, Mississippi State University

Inscription on Darden Monument, Mississippi State University

Lime Rock Grange, Lincoln, Rhode Island

Iowa Grange Halls

Grange Memorial Dormitory, Pennsylvania State University

Mary Mayo Hall, Michigan State University

pomona cooperatives and agencies that beset state-run enterprises. There were dishonest agents who speculated with chapter funds and others who took order money and visited "bunko and faro shops." There were also operating difficulties. In 1875 Illinois State Grange Agent S. J. Frew investigated local and county cooperative businesses and then summarized candidly his detailed findings in a special report delivered before the state grange. Frew charged that these enterprises were too independent of his agency and complained that they aided monopolies by "supply[ing] their stock of goods from all sources, irrespective of the Rings that [are] our avowed enemies." In addition, he surmised that no advantages accrued from dealings with these cooperative firms because their prices were no lower than those found at neighborhood retail markets. In conclusion, because of these factors, Frew judged these efforts "failures." [46]

Coupled with their other programs, granges also operated their own banks. In 1875 the Virginia legislature issued charters for the Grangers' Banking and Trust Company of Fredericksburg, the Staunton River Grange Savings Bank of Pittsylvania County, and the Border Grange Bank. Outside Virginia, calls were made for the establishment of state grange banks, but nothing resulted from these pleas. A most elaborate scheme for a grange bank was never tested. In late 1873 Master S. W. Land of Mississippi's Rocky Point Grange No. 104 laid out complete plans in the *Farmers' Vindicator* for a state grange bank. Land suggested that every chapter should be enlisted to contribute twenty-five 450-pound bales of cotton to the state grange. Proceeds from a sale of cotton wuold be used to purchase gold reserves. Thereupon, the bank would issue its own notes and liquidate its debts for the cotton. Land noted two advantages of such a bank. It would be able to make low-interest loans, and it could prevent foreclosures of members' properties by lending them money for paying off mortgages held by other banks. [47]

Providing protection for members against the unexpectant perils of fire and death was another facet of the grange business program. A combination of factors drove granges into the insurance field. In the first place, the order promoted brotherhood, and paternalism was a logical result of fraternalism. Second, there was a growing belief

46 Macon *Missouri Granger*, March 30, 1875; Friendship, Wisconsin, *Adams County Press*, May 23, 1874; Illinois State Grange, *Proceedings*, 1875, pp. 24–31.

47 Virginia *House Journal* (1875), 171, 178, 235, 253, 264, 299; Virginia *Senate Journal* (1875), 208, 217, 235, 250, 309, 318–19; South Carolina State Grange, *Proceedings*, 1875, n.p.; Charleston *Rural Carolinian*, January, 1874.

among farmers in the late nineteenth century that private insurance companies were levying "exhorbitant [sic] rates" for protection.[48]

Grange mutual insurance companies fell into two major classifications—those paying indemnities for deaths and those covering fire losses. Each operated on the same basic principles. Eligibility for benefits was dependent upon fulfilling two obligations. Insurants had to keep up their membership in the Grange, and they had to pay assessments and initiation fees. Premiums fluctuated in direct proportion to the number of legitimate claims. If, for example, three insurants died during a twelve-month period, assessments for paying off the deceased persons' beneficiaries were levied three times.[49]

Fixed criteria determined how much insurants had to pay for reported claims. Mutual fire protection associations based their assessments almost entirely on the amounts of coverage carried by individual policyholders. On the other hand, grange life insurance companies used members' ages and their desired indemnities to determine amounts levied. If an insurant were thirty-three years old and covered for $2,000, he paid greater assessments than an individual who was 24 with a $1,000 indemnity.[50] The following table shows how the Ohio State Grange Mutual Protection Association classified its participants and how it assessed members of each category for every reported claim:

Maximum assessments for members having $1,000 coverage:
Members, 20–35 (Group A), paid $1.00 per death.
     "   , 35–50 (Group B),   "   1.25   "   "   .
     "   , 50–65 (Group C),   "   1.50   "   "   .

Maximum assessments for members having $2,000 coverage:
Members of Group A paid $2.00 per death.
     "   "   "   B   "   2.50   "   "   .
     "   "   "   C   "   3.00   "   "   .

[48] Iowa State Grange, *Proceedings*, 1873, p. 40. See also: Indianapolis *Indiana Farmer*, February 12, 1881; New Hampshire State Grange, *Proceedings*, 1876, pp. 14–17; Missouri State Grange, *Proceedings*, 1889, p. 16; Patrons' Benevolent Aid Society of Wisconsin, *Second Annual Circular* (Madison, 1878), 1.

[49] Aid Society of Wisconsin, *Second Annual Circular*, 2–3; Patrons Aid Association, Orangeburg, S.C., *By-Laws* (Charleston, 1877), 9, 12–13; Oregon State Grange, *Proceedings*, 1889, p. 16; Massachusetts State Grange, *Proceedings*, 1886, p. 47; 1887, p. 18; Connecticut State Grange, *Proceedings*, 1887, pp. 10–11.

[50] Aid Society of Wisconsin, *Second Annual Circular*, 2; Patrons Aid Association, Orangeburg, *By-Laws*, 12–13.

Maximum assessments for members having $5,000 coverage:
Members of Group A paid $5.00 per death.
     ”   ”   ” B   ”   6.25 ”   ” .
     ”   ”   ” C   ”   7.50 ”   ” .[51]

Mutual insurance associations succeeded in reducing protection costs; in fact, mutual assessments were usually much less than premiums charged by private companies. Statistics released by state insurance commissioners revealed that rates levied by stock companies were often two and three times as much as assessments paid by mutual insurants. Of course, state grange committees on insurance used the data provided by state commissioners to put together publicity announcements which emphasized the great insurance bargains offered by mutuals. One such release issued by a New York committee in 1881 alleged that the state's cooperative companies had covered properties in 1880 for $57,471 less than stock companies would have charged for equivalent protection.[52]

Granges provided insurance services during the First Granger Movement, but these efforts were not in any way comparable with those extended during the Second Granger Movement. Throughout the last two decades of the nineteenth century, farmers turned increasingly to the Grange for protection. As a result, the number of mutual insurance companies grew steadily, and the total amount of property insured by grange companies doubled several times. The turn to granges for insurance protection generally paralleled membership gains made by the order in the Northeast during the last two decades of the nineteenth century. Growth of grange mutual insurance companies in New York and New Hampshire certainly followed on the heels of the order's resurgence in these states. While the organization was experiencing a popularity revival in New York, the number of grange fire relief associations in the state increased from fifteen in 1880 to seventy-two in 1891 and the value of risks covered by these companies soared from $9,177,481 to $106,895,314.[53] New Hampshire witnessed a similar phenomenon at the end of the century.

The most ambitious scheme for cooperation ever offered to the

51 Ohio State Grange, Mutual Protection Association of the Patrons of Husbandry, By-Laws (Kenton, Ohio, 1876), 10–12, 17–18.

52 New York State Grange, Proceedings, 1880, pp. 75–76; 1881, pp. 34–35.

53 Ibid., 1880, pp. 19–20; 1891, p. 119; New Hampshire State Grange, Proceedings, 1890, pp. 16–17; 1899, p. 8. Check Table 10 on page 43. New York State Grange, Proceedings, 1881, pp. 34–35; 1892, p. 73.

Grange for adoption never fully got off the ground. In 1873 Dr. Thomas Worrall apparently conceived of the Mississippi Valley Trading Company (MVTCo.) as an instrument of direct trade between grange cooperatives and English Rochdale associations. Worrall was from Great Britain, but he had worked as a resident businessman in New Orleans since the Civil War. The executive committee of the National Grange first considered and became actively involved in matters of the MVTCo. during the summer of 1875.[54]

MVTCo. organizational framework was to consist of two self-sustaining bodies. One was to be English, while the other was to be American. Each was to issue 100,000 shares of stock either at one pound or at $5 per certificate, and both sections were to have their own board of trustees. Overall direction for the MVTCo. was to be provided by a council composed of two boards meeting together. Worrall's plan for the MVTCo. was pinned on the removal of three tolls affecting American trade with Great Britain. Worrall saw commerce between the two nations suffering from commissions to English merchants for goods sold in the United States, from commissions to factors who bought commodities for consignment abroad, and from burdensome railroad freights between the American heartland and eastern ports. The Englishman believed that a union of British and American cooperatives would eliminate middlemen's profits because cooperatives would handle their own shipping and distributing. Solution to the railroad-tariff problem involved shifting commerce from its west-east axis to a north-south course along the Mississippi River. By this change, Gulf ports would be serving as the major exchange and storage points for transoceanic transport. [55]

[54] A special expression of gratitude must be given to Dr. Howard B. Schonberger, a former graduate student at the University of Wisconsin and a professor of history at Hampton Institute, for unselfishly and willingly calling the author's attention to several sources used for the development of the MVTCo. Some of Schonberger's research appears in: *Transportation to the Seaboard: The "Communication Revolution" and American Foreign Policy 1860–1900* (Westport, 1971). Mississippi Valley Trading Company, Ltd. Papers, edited by Charlotte J. Erikson, Library of the Cooperative Union, Manchester, England. Papers are on microfilm at the Minnesota State Historical Society Library, St. Paul. Information for this paragraph was based upon the introduction by Erikson.

[55] Mississippi Valley Trading Company, Ltd., *Memorandum and Articles of Association* (Manchester, England, 1875); Clifton K. Yearley, Jr., *Britons in American Labor: A History of the Influence of the United Kingdom Immigrants on American Labor, 1820–1914* (Baltimore, 1957), 243–57. Yearley not only highlights the history of the MVTCo. in the United States, but he also traces the development of Grange-operated cooperative enterprises and concludes that the genesis of these businesses

Confident that the MVTCo. would gain acceptance, Worrall left for Great Britain late in 1873 to speak in behalf of his proposal before chambers of commerce and Rochdale cooperatives. While in England, Worrall enlisted support of the hierarchy of the English Cooperative Union, a loose federation of industrial cooperative societies. The union's leader, Edward V. Neale, immediately saw great possibilities accruing from direct trade. It would open new markets for excess productions of British worker-owned industries as well as provide employment for England's jobless. In addition, it might strengthen the cooperative movements of the two nations.[56]

Upon his return from England in the spring of 1874, Worrall commenced a campaign aimed at inducing grangers to support the MVTCo. One of his first stops was Albany, Georgia, where the Direct Trade Union of the Patrons of Husbandry (DTU) was having an organizational meeting. Worrall actively sought the assistance of the DTU. The Georgia State Grange had organized the DTU in June, 1874, as a direct trade line between southern cotton farmers and British textile manufacturers. The DTU had wide support in the South and employed agents in Liverpool, England; Norfolk, Virginia; Charlotte, New Bern, and Raleigh, North Carolina; Madison, Conyers, Savannah, Griffin, Macon, Augusta, and Rome, Georgia; and Mobile, Huntsville, and Decatur, Alabama.[57]

Worrall tried to show leaders of the DTU the similarity between their organization's objects and those of the MVTCo. According to the DTU secretary, the transplanted Englishman "depicted most truthfully & powerfully the dependent & poverty-stricken condition of the Southern States—their thraldom to the great cities & money-power of the New England States" and offered his scheme for direct trade with English industrial cooperatives. The MVTCo., noted Worrall, presented southern and western farmers and British industries with the opportunity of being "brought in direct contact without

came from Britons within the order. See also Philip N. Backstrom, "The Mississippi Valley Trading Company: A Venture in International Cooperation 1875–1877," *Agricultural History*, XLVI (July, 1972), 425–37.

56 Introduction by Erikson, MVTCo. Papers; Beatrice Potter (Webb), *The Cooperative Movement in Great Britain* (London, 1930), 172–75.

57 E. T. Paine, *The Direct Trade Union: Its Objects and Advantages* (Atlanta, 1874), 1–16; Tennessee State Grange, *Proceedings*, 1875, pp. 38–39; Alabama State Grange, *Proceedings*, 1874, p. 22; South Carolina State Grange, *Proceedings*, 1875, n.p.; Atlanta *Wilson's Herald of Health, and Farm and Household Help*, December, 1873; Direct Trade Union, Circular by A. H. Colquitt, *To the Patrons of Husbandry of the Cotton States* (Atlanta, 1875).

the intervention (as now) of the middlemen of New York and other Northern cities."[58]

Although the Grange never gave complete approval to Worrall's plan, he was satisfied that this would come as soon as he coaxed the English Cooperative Union to support his plan. Worrall returned to Great Britain for a second time in 1875 to secure English approval. His mission involved persuading Rochdale industries to join the MVTCo. To succeed, Worrall had to overcome three formidable administrative hurdles. Before proceeding, he had to obtain consent of the divisions of the central board of the English Cooperative Union, then approval of the associations which they represented, and finally endorsement of the whole body at the annual congress of cooperative societies. Nevertheless, by April 15, 1875, Worrall had cleared the last obstacle blocking English acceptance of the MVTCo.[59] The next step involved drafting articles of incorporation for the British sector of the MVTCo. and selecting officers. Worrall worked closely with leaders of the English Cooperative Union to draw up a constitution. Its by-laws explained the proposed company's purpose and laid down structural guidelines. As soon as the draft had been completed and approved, officers were named in accordance with articles. Worrall was chosen managing director of the MVTCo.[60]

After the British had formed their section of the MVTCo., comparable actions had to be taken in the United States by the Grange in order to complete MVTCo. preliminaries. Prodding grangers to organize their sector became a special charge of Secretary Neale of the congressional board of the English Cooperative Union. He circulated letters to national, state, and subordinate grange officers and members, urging them to consider advantages of the MVTCo. and encouraging them to organize the American counterpart to the British section.[61]

Sensing that letters were not exerting enough pressure on granges, stockholders in the British sector dispatched a five-member deputation to the United States July 10, 1875, to assist in the formation of an American half of the MVTCo. As things turned out, the Englishmen's

---

[58] Direct Trade Union, Secretary's Minutes, September 20, 1874, in Branch Family Papers.

[59] Edward V. Neale to Officers and Members of the National, State, and Local Granges of the Order of Patrons of Husbandry in the Mississippi Valley, April 15, 1875, in Osborn Papers.

[60] *Ibid.*, Worrall to John Neall, April 27, 1875, Worrall to H. R. Bailey, April 29, 1875, both in MVTCo. Papers.

[61] Neale to Officers and Members of the National, State, and Local Granges, April 15, 1875, in Osborn Papers.

efforts were not altogether needed. By the time the British delegates arrived at New Orleans, favorable actions had already been taken in behalf of the MCTCo. by the executive committee of the National Grange. The grange council had endorsed and recommended the direct trade scheme to state as well as local granges for their adoption.[62] Although the executive committee had given its sanction to the MVTCo., the Britons proceeded with their assigned task of drumming up support for the direct trade proposal. Upon their arrival in the United States, these deputies split up and went in different directions to cover as much territory as possible. Each delegate spoke to several granges of the bright trading prospects to be offered in the future by the MVTCo. After these representatives of the English Cooperative Union had felt assured that they had created enough grass-roots support for the MVTCo. to guarantee formation of the granger section, they began converging at St. Louis for a final meeting with the National Grange executive committee. The Englishmen found the mood at the session to be one of friendliness and warmth. From this congeniality, they surmised that grangers would act favorably on the MVTCo. as soon as such action was possible. Satisfied with their mission and its results, the deputation entrusted Worrall with responsibility of bringing grangers into the MVTCo. and then embarked for Europe.[63]

Records for the next few months are sketchy. From all available evidence, it appears that Worrall and three associates acted on their own, without consultations and without sanctions, to organize the American Cooperative Union at Louisville, Kentucky. Tabbed as the wholesaling adjunct of the MVTCo. by Worrall in a tract, the union was to contribute one-fifth of its stock to the MVTCo. and was to sell remaining certificates on the open market.[64]

Angered by openness, members of the National Grange executive committee reacted decisively against the American Cooperative Union's provision which permitted nongranger participation. Worrall and his three colleagues were charged with undermining the goal of international cooperation. Prospect of unrestricted membership greatly disturbed grange officials; they feared it would enable middle-

[62] Worrall to John Cochrane, July 21, 1875, in Osborn Papers; Dr. John H. Rutherford to the Cooperative Societies of the United Kingdom, July 10, 1875, in MVTCo. Papers.
[63] National Grange, *Proceedings*, November, 1875, pp. 89–90; 1876, pp. 48–67.
[64] American Cooperative Union, *Manual of Practical Cooperation by Thomas D. Worrall, Louisville, 1875* (Louisville, 1875).

men and other enemies of the order to step in and subvert the program. Thus, when the whole question of grange involvement came up for inquiry in October, 1875, the National Grange executive committee ruled against affiliating with the MVTCo. until assurances were given that the articles of association would be revised to grange specifications.[65]

Meanwhile, English officers of the MVTCo. responded with concern; they recognized at once what the executive committee-Worrall dispute meant. Since they knew well how important the Grange was to the overall scheme, they did not want any unnecessary rifts over technicalities spoiling the whole direct trade program. Consequently, Secretary Joseph Smith and Director Neale each wrote Worrall urging him to retreat from his adamant position.[66]

Soon, some encouragement came from across the Atlantic. Master John T. Jones of the National Grange informed the British sector of the MVTCo. that Worrall's rift with the executive committee had not choked off the order's desire for business relations with English cooperatives. But he warned that certain problems had to be ironed out before the organization would give its final clearance for commencing operations. Jones was particularly concerned about finding sources of credit. Forty years experience as an Arkansas cotton grower had taught him the importance of monetary advances to farmers, so he cautioned British directors that the size and success of the MVTCo. might very well depend on the amount of lending power it possessed.[67]

Prospects still appeared bright for success when the National Grange convened at Louisville in November, 1875, for it soon became evident that the Grange's conflict with Worrall had as yet not affected the order's desire to participate in the MVTCo. In fact, interest in direct trade with English cooperative societies was manifested throughout the session. A climax came when the National Grange resolved that J. W. A. Wright be commissioned to serve as its special envoy in Germany and the United Kingdom. While in Europe, Wright was to pursue possibilities of renegotiating conditions for the American sector of the MVTCo. and was to ascertain whether Germans and Britons were interested in forming granges. Of course, Wright's appointment

65 National Grange, *Proceedings*, November, 1875, pp. 91–93.
66 Neale to Worrall, January 25, 1876, Joseph Smith to Worrall, October 28, 1875, both in MVTCo. Papers.
67 National Grange, *Proceedings*, November, 1875, pp. 91–93.

was a slap at Worrall because it meant that the order no longer placed any confidence in his efforts to develop an American branch for the MVTCo.[68]

For Worrall, this was no small matter; he did not casually shrug off the danger which Wright's mission posed to his position in the MVTCo. Instead, he countered with a risky strategy aimed at cutting away rank-and-file grangers' trust in their leaders. By fomenting dissension, Worrall hoped to restore his prestige. He expected a vacuum to develop, from which he would emerge as the unquestioned spokesman for the MVTCo. Worrall also took his campaign to England. On January 21, 1876, he wrote to Neale, advising him to "tell Mr. Wright that if he wants any thing or if the National Grange Executive [Committee] want any thing with you, you have a duly accredited agent in the United States who being on the spot is in the best possible position to inform the board." The propaganda campaign failed; the English cooperative societies' leaders gave Wright a friendly reception upon his arrival February 24, 1876, and grangers never disavowed their officers.[69]

This is not to suggest that Wright's arrival in Great Britain had not placed the English directors of the MVTCo. in an awkward position. If they welcomed the grange representative, they would be risking an open break with Worrall. On the other hand, if Wright were ignored, all opportunities for direct trade with grangers might be irrevocably ended. Although the ultimate decision of what to do under the circumstances no doubt had not been an easy one to reach, the Englishmen finally decided that practical considerations were worth more in the future than sentimental attachments to the man who was the MVTCo.'s founder, its managing director, and its chief advocate.[70]

After exploring trade possibilities in Germany and determining the reaction of Germans to being grangers, Wright investigated operations of several cooperative societies in England before finally going to Manchester to begin negotiations for new terms for the MVTCo. His mission was purely exploratory as he had been instructed not to make any commitments. Wright was merely supposed to feel the En-

[68] *Ibid.*, 89–93; 1876, pp. 48–67; Charleston *Rural Carolinian*, February, 1876.
[69] Louisville *Agricultural Economist*, February 1, 1876, clipping in MVTCo. Papers; Louisville *Jeffersonian Democrat*, February 5, 1876, clipping in MVTCo. Papers; National Grange, *Proceedings*, 1876, pp. 48–67; Worrall to Neale, January 21, 1876, in MVTCo. Papers.
[70] Neale to Worrall, January 25, 1876, in MVTCo. Papers.

glishmen out in order to see what concessions the Grange might be able to wring from ensuing discussions.[71]

Negotiations had no sooner begun than Wright discovered a general unwillingness on the part of his English hosts to break clearly with Worrall. The "ambassador" confided that the order was prepared to bring outsiders into the MVTCo. as long as there were guarantees that they would not be in positions to tamper in any way with the inner workings of the farmers' organization. Wright emphasized that what disturbed grangers was not what Worrall advocated but the way he and his associates tried to appeal to grangers over the heads of their leaders.[72] Wright's mission was successful in that it forged three major changes in the MVTCo.'s articles of association which had been desired by the Grange. The name, MVTCo., was dropped for the Anglo-American Cooperative Trading Company; the position of managing director was abolished; and the right to purchase stock in the company was restricted to cooperatives. By accepting these alterations, British directors inflicted a stunning defeat on Worrall. Amendments to the articles of association limited stockholding to participating associations, blocking involvement by bodies like Worrall's American Cooperative Union; they eliminated the post that he occupied; and they broadened geographic scope of the American branch, thereby lessening Worrall's chances for a comeback. By making direct trade a national project, grange leaders reduced the influence of the lower Mississippi Valley, the heartland of Worrall's supporters.[73]

When news of the amendments to the articles of association reached Worrall, he simply ignored them, continuing his efforts in behalf of the American Cooperative Union as if nothing had happened. He still dreamed of a day he personally would supervise a massive direct trade network between the Mississippi Valley and English cooperative societies. Grangers were dismissed as a bloc of individuals guided by selfish leaders dedicated to their own ends. Despite their determination, Worrall and his Louisville associates failed to move English Rochdale leaders to conduct trade on a separate basis with the farmers' order and themselves. After months of struggling, Worrall finally aban-

---

[71] National Grange, *Proceedings*, 1876, pp. 48–67.
[72] *Ibid.*; John T. Jones to J. W. A. Wright, April 8, 1876, in MVTCo. Papers.
[73] Charleston *Rural Carolinian*, September, 1876; Executive Committee of the MVTCo., Ltd., Changes in the Articles of Association, Manchester, May, 1876, in MVTCo. Papers.

doned his campaign for recognition of the American Cooperative Union in August, 1876.[74]

Meanwhile, Wright was submitting plans for the new Anglo-American Company to members of the executive committee and to the National Grange for their consideration. The Wright mission's obvious success did not pass unnoticed. Upon reviewing the new articles of association, Master Jones projected that "the time has arrived [to] take some definite action." At the 1876 session, the head of the National Grange devoted most of his annual address to the need for international cooperation and to the formation of the Anglo-American Cooperative Trading Company, while Wright spoke of his efforts and told of the British cooperators' desire for prompt and definite action. Therefore, responsibility for examining the whole matter and reporting on it was delegated to a three-man committee on cooperation. This body was composed of Masters Thomas R. Allen of Missouri, E. R. Shankland of Iowa, and Joseph H. Osborn of Wisconsin.[75] Their report endorsed unequivocally the Anglo-American Cooperative Trading Company. Besides explaining their decision to sanction the plan, their study also laid down six recommendations. First, the committeemen suggested that British cooperators be thanked for their steadfastness and understanding. Second, they proposed that the National Grange "accept the general plan . . . to form . . . as soon as practicable the . . . Company, for the purposes named in their memorandum and articles of association as amended." Third, they advised members "to be prepared to take part of the proposed American shares in this company." Fourth, they recommended that English directors be notified by the master and executive committee of the National Grange as soon as they have learned of the willingness of grange cooperative associations "to subscribe the $125,000 gold of American shares of stock of this company to establish the necessary shipping and receiving depots, and managers in our American seaports." Fifth, as a preliminary to formal acceptance, Cochrane, Shankland, and Allen asked that eager grange associations that did not want to wait for formal approval of the Anglo-American Company be permitted to begin negotiations immediately with Vansittart Neale to ascertain

---

[74] American Section of the MVTCo., Minutes of the Executive Committee Meetings, June 15, 17, 1876, in MVTCo. Papers; Louisville Cooperative Journal of Progress, June 1, 1876, clipping in MVTCo. Papers.

[75] National Grange, Proceedings, 1876, pp. 8–10, 48–67, 124–30.

whether any English cooperative societies wished "to buy directly
from our associations, the wheat, cotton, corn, wool, cheese, pork, tobacco, and other farm products needed for their consumption." The
final suggestion filed by the three Masters provided a scheme for in-
stallment buying of stock. With a few minor revisions, the National
Grange approved the committee's resolutions.[76]

Endorsement of the executive committee and the National Grange
did not come without opposition. Three delegates attending the 1876
conclave blasted the entire effort. After labeling the proposal "a
scheme of business entirely unsuited to the wants and conditions of
American farmers," the three skeptical critics suggested that "it is the
better policy for our farmers to buy and sell as near home as practicable." Their fight brought a motion postponing action indefinitely on
resolutions favoring the Anglo-American Company to the floor for a
vote, but it was defeated twenty-three to sixteen.[77]

Positive actions taken by the National Grange at its 1876 session
made international cooperation appear close at hand. Unfortunately,
that hope soon vanished because of a worsening of economic conditions in England and because of a foolish investment there. With each
passing day, the British economy slipped further into depression. As
a result, thoughts turned increasingly from the far-distant future to
day-to-day tasks of providing necessities, while capital once available
for investments such as the Anglo-American Company disappeared.
An unhealthy English economy contributed much to the downfall of
international cooperation, but a more serious blow came when the
Leeds Cooperative Society, holder of two thousand of the forty-two
hundred subscribed shares, withdrew from the program because of a
loss sustained by investing heavily in a fraudulent coal-mining operation. This withdrawal prompted British directors to appoint a liqui-
dator January 24, 1878, to dissolve the Anglo-American Trading Co.
Ltd.[78]

Meanwhile in the United States, interest of grangers in the project
had been sufficient for its success, but the number of stock subscriptions had not been. Tumbling memberships and folding enterprises
were making grangers more and more hesitant to take risks. By the
time calls were issued for stock subscriptions to the Anglo-American

---

[76] *Ibid.*, 148–50, 167–68.   [77] *Ibid.*, 144–52.
[78] Neale to Wright, July 10, September 26, 1877, Neale to National Grange Executive Committee, January 3, 1877, unsigned note in the Anglo-American Cooperative Trading Co., Ltd. Letterbook, January 24, 1878, all in MVTCo. Papers.

Company, members' fresh memories of previous failures warned them against sustaining additional losses. The result was that not enough grangers agreed to invest in the Anglo-American Company to stake the unproven venture.[79]

For many grangers, the Anglo-American Trading Company was just another futile attempt at direct producer-to-consumer international trading, and they wanted no part of it. Southern members were skeptical because of the failure of the DTU of Georgia,[80] while West Coast patrons doubted whether granger schemes for direct trade would work because of their knowledge of largely unsuccessful efforts of western wheat growers who had tried to bypass normal trade channels.

Grangers from the San Francisco Bay area had declared war against Isaac Friedlander in 1874. Dubbed the "Grain King" by his contemporaries, Friedlander had reigned as the leading wheat purchaser and speculator in San Francisco since the Civil War, and grangers plotted to take away his crown. They combined and dispatched ships laden with grain for the ports of the world. In 1876 three ships—the *Star of Hope*, the *W. R. Grace*, and the *Seaton*—loaded with patrons' wheat sailed from California for Liverpool. These voyages brought a total return of $1,172,439.26, but from this amount the recipients had to pay $27,653.30, in commissions. Although receipts from these granger shipments were impressive, such trade still was unable to compete with Friedlander because he bought on consignment and extended credit to farmers. As a result, the granger attempt to destroy the Grain King's empire fizzled.[81]

Mississippi grangers used a somewhat different approach to secure a foreign market for their cotton. In 1873 the state grange employed at Liverpool a resident agent who served as the chapter's special broker. Under the system, members' cotton was collected and shipped to him. Thereupon, he bargained with buyers to obtain the highest price available. The arrangement had merit, but once again most grangers were not able to take advantage of a good plan because their old nemesis, credit, obligated them to the old system. When the pro-

[79] National Grange, *Proceedings*, 1877, pp. 12–13.
[80] See footnote 57.
[81] Charleston *Rural Carolinian*, November, 1874; Chicago *Prairie Farmer*, September 30, 1876; Rodman W. Paul, "The Great California Grain War: The Grangers Challenge the Wheat King," *Pacific Historical Review*, XXVII (November, 1958), 331–49; Grangers Business Association, San Francisco, Aggregate Sales and Commissions, February 1, 1876–January 1, 1877, in MVTCo. Papers.

gram was inaugurated, most of the next crop had already been assigned to merchants as payment for debts. Conditions being what they were, the chapter's Liverpool representative soon found that there was not enough business to justify his position, so he returned home.[82]

The existence of so many plans for cooperation suggests something about business attitudes held by members of the farmers' order. Many grangers bitterly opposed the capitalist system as it existed in late nineteenth-century America. In particular, they disliked trusts, monopolies, and shylocks. Master Milton Trusler of Indiana was a leading spokesman of this faction. He believed class conflict was inevitable in the United States unless some sweeping governmental proposals for redistributing wealth were inaugurated soon. A clash would come because poverty and hunger were pushing lower class citizens to desperation while increasing wealth was making rich people more and more aristocratic, dominant, and oppressive. He warned that the two elements in society could not be reconciled with present-day approaches because the two are "as foreign as liberty and slavery." Trusler advised his countrymen to do some serious soul searching because "there must be something radically wrong" when a few men could earn millions, while others hovered in poverty.[83]

To correct these problems created by concentrated wealth, granges recommended structural changes in the nation's economy. A five-point program suggested at the turn of the century by the Indiana State Grange was typical of these recommendations. In essence, it called for official inspection of corporations' books, prohibition of all rebates and discriminations by public carriers, taxation of all capital stock, collection of taxes on all capital stock certificates, and severe penalties for all law violators.[84]

Others in the Grange admired Horatio-Alger-type plutocrats, and frowned on changes. On October 18, 1893, members of Social Grange No. 138 applauded "an interesting talk regarding the millionaires of the country to the effect that all did not gain their wealth by unjust means" because they cherished opportunities offered Americans to

[82] James S. Ferguson, "Co-operative Activity of the Grange in Mississippi," *Journal of Southern History*, IV (January, 1942), 10.

[83] Pennsylvania State Grange, *Proceedings*, 1887, pp. 16–21; Indiana State Grange, *Proceedings*, 1898, pp. 9–10; New York State Grange, *Proceedings*, 1891, pp. 124–25; Illinois State Grange, *Proceedings*, 1891, pp. 70–71; Indiana State Grange, *Proceedings*, 1891, p. 6.

[84] Indiana State Grange, *Proceedings*, 1899, p. 11.

make money and elevate themselves.[85] Among conservatives in the order, Master Edward Wiggins of Maine had no peer. His addresses hammered home the theme that no one benefited from class consciousness:

> Conservative men in our order unite in discouraging any antagonism between capital and labor, recognizing the fact that the interests of each are identical. The shout against combined wealth and the power of corporations is often but the demagogue's cry for the purpose of furthering his own ambitious designs. Without large aggregations of capital no important enterprises can be carried on, and with the discontinuance of these enterprises labor languishes and seeks employment in vain. It is only when those who control capital combine to oppress the people and defraud them of their rights that wealth becomes the enemy of labor.[86]

If grangers had been polled on economic questions and policies, results would have shown divergent opinions and responses. Nevertheless, the wide assortment of grange cooperative enterprises reflected general antibusiness attitudes. Most grangers distrusted entrepreneurs, and their suspicions engendered programs aimed at breaking up trusts and monopolies and punishing thieves and swindlers. In effect, the order tried to become a watchdog against fraud, deceit, and theft.[87]

Although the order was certainly committed to fight fraud and deceit, grange organs still were not very selective when they solicited advertisements. Toadstool medicine men filled order journals with fraudulent claims. In one issue of the *Georgia Grange*, unsuspecting farmers were told that dosages of Simmons Liver Regulator cured dyspepsia, constipation, jaundice, bilious attacks, headaches, colic, sour stomachs, heartburn, and depression of spirits. In the same magazine, patrons were advised to take Dr. Bond's Discovery for cancer because it eliminated the malady with "no knife, no caustic, no plasters, no pain." On occasion, grange journals did guard their readers from dishonest product claims by refusing to publish certain advertisements. For example, the *Patron of Husbandry* of Columbus, Mississippi, did not print solicitations from "disreputable men in the cities." [88]

[85] Social Grange, Secretary's Minutes, October 18, 1893.

[86] Maine State Grange, *Proceedings*, 1897, p. 13.

[87] Wisconsin State Grange, *Proceedings*, December, 1890, p. 7; Macon *Missouri Granger*, editorial policy.

[88] Atlanta *Georgia Grange*, July 14, 1877; Columbus (Miss.) *Patron of Husbandry*, July 26, 1879.

Although grangers did not necessarily show respect for corporate holdings, they wanted their own possessions protected. In areas where law enforcement was lax, granges formed vigilante and crime detection committees to protect members' property. Granger policing was very common in former slave states. Many southern whites feared that the section's socioeconomic turbulence and its shortage of law officers were increasing possibilities of criminal activity in the region. As these fears were sweeping through the South after the Civil War, granges often responded to them by forming vigilante groups to protect members' possessions from thieves.[89]

By bringing farmers together at least once a month, granges were also able to warn their members and others about swindlers. Late in 1873 the Kenoma Grange of Anderson County, Kansas, showed how the order was in a good position to protect rural residents. A report came to this Kansas subordinate "that bonds of Anderson County, issued in aid of the PAOLA AND FALL RIVER RAILROAD COMPANY, are fraudulent, and the payment thereof will be resisted to the last," so members of the local quickly circulated handbills around the county, cautioning citizens not to invest in these certificates until after the rumor had been investigated and discredited.[90]

There were many factors attesting to the fact that grange business programs had an impact on rural America. Nongrangers seeking a testing place for their panaceas submitted plans to granges for their consideration, and critics found it necessary to lampoon the organization's efforts. Some of their criticisms were very clever. *Western Independent* printed a sarcastic story of a patron who dreamed he had died and had gone directly to the "spirit world," where he soon found that neither heaven nor hell wanted him because of his affiliation with the secret farmers' order. "After he had gone way off [in hell,] he was accosted by the homely ruler of the pit, when [*sic*] the following propositions were made: 'Stranger,' said Nick, 'I will not admit you here; they do not want you in heaven; but I will send you two hundred barrels of brimstone for cash, then ten per cent off, and you can start a little hell of your own, with no agents or middlemen.' "[91]

Evidence suggests that retail grocers felt some encroachment of

[89] Topeka *Kansas Farmer*, September 30, 1874; Chicago *Prairie Farmer*, June 16, 1877; Louisville *Farmers Home Journal*, September 4, 1879.

[90] Kenoma Grange, Anderson County, Kansas, Circular entitled *Be Warned*, December 22, 1873.

[91] Thomas J. Durant to the National Grange Executive Committee, May 6, 1874, in Saunders Collection; Fort Smith, Arkansas, *Western Independent*, June 25, 1874.

grange cooperative stores. That they minced in face of challenging granger competition was seen from several sides. The pages of the *American Grocer* tell of retailers' concern. On one occasion, the editor wrote an article telling private storeowners how to compete with cooperatives. The editorial advised grocery dealers to "get down to a strictly cash basis" because "if you do that, you have every advantage in your favor, for you are better buyers and better business men." [92]

On another plane, a poll taken by the New York State Grange in 1877 supplied some rather conclusive evidence concerning the order's claim that it was affecting members' buying habits. To the question— "Have the members of your Grange received pecuniary benefits on account of their connection with the order?" almost every respondent replied in the affirmative. Many replies noted particularly high savings on purchases of household supplies.[93]

Cooperation played a decisive role in the order's program; yet, granger spokesmen cautioned members not to give it priority. Overseer Ira McCallister of Nebraska summed up feelings of many on this whole subject when he reaffirmed the contention that the "business features of our work . . . [must be given only] secondary consideration." [94] In summary, therefore, grange business operations were important, but they were not the main reason for the order's existence.

92 Topeka *Kansas Farmer*, April 12, 1876.
93 New York State Grange, *Proceedings*, 1877, p. 78.
94 Ohio State Grange, *Proceedings*, 1898, p. 13; Wisconsin State Grange, *Proceedings*, 1879, pp. 19–20; Nebraska State Grange, General Records of the Secretary, December, 1887.

CHAPTER VIII

The Grange
and Partisan Politics

A thin line separates partisan politics from political involvement. Grange bylaws denied chapters the right to endorse candidates and forbade members to run for public office under the banner of the order, yet grangers were urged to work for legislative solutions to agrarian problems. The upshot was a constant dilemma. Patrons found themselves asking what steps could be taken to push legislation through state assemblies and Congress, and which could not be used.[1]

Minnesota grangers were among many who faced this problem in the 1870s. The result was a state-wide rift which left the order in splinters. The dispute began January 30, 1873, when an ambitious second-generation Irish-American named Ignatius Donnelly joined Cereal Grange No. 25 at Hastings. From the very beginning, Donnelly demonstrated what he expected from his membership. He viewed it as an opportunity for launching a new political career. The day he affiliated with the Grange, his brothers and sisters elected him chapter treasurer. Since the position gave its occupant chances to show his speaking abilities and inadvertently extended to him opportunities to tell others of his qualifications for political office, Donnelly accepted the post.[2]

Throughout the spring, Donnelly told grangers that Democrats and Republicans had refused to work for solutions to agrarian grievances. As he spoke, news spread throughout Minnesota of a rebel in the order's ranks, one who was challenging the state's corrupt and inept two-party system. Within a short time, granges throughout Minnesota were flooding the insurgent with invitations to speak. So great was the response that Donnelly noted in his diary, "the farmers are evi-

---

[1] New York *Times*, February 7, 1874; Ohio State Grange, *Proceedings*, 1893, p. 6.

[2] Ignatius Donnelly Diary, January 30, 1873, in Ignatius Donnelly Papers, Minnesota State Historical Society Library, St. Paul; Ridge, *Ignatius Donnelly*, 149–61; Ridge, "Ignatius Donnelly and the Granger Movement in Minnesota," 693–709.

dently a good deal roused up about the questions which concern their mutual interests." [3]

When grangers from Rice, Dakota, and Goodhue counties gathered at Northfield on June 17, 1873, for a day of relaxation, Donnelly was on hand to speak. He used the opportunity to deny published reports carried by the St. Paul *Press* and the Minnesota *Tribune* that accused him of being an active political officeseeker. The news stories, said Donnelly, were rumors which should be squelched.[4]

Nevertheless, Donnelly was creating grass-roots support. His followers praised his efforts. Requests from granges wishing to hear him increased, and letters praising his activities and calling on him to organize a farmers' party came in greater numbers. On July 25, 1873, Editor Horace White of the Chicago *Tribune* wrote Donnelly to inform him of third-party movements in Iowa and Wisconsin. White advised the Minnesotan to wait before launching a party because "it was too early to hold a general anti-monopoly convention." According to the newspapermen, farmers needed "to consolidate and strengthen their organization" before taking a plunge into the political arena.[5]

After working rigorously for several months, Donnelly sensed that he had drummed up enough support to form a successful third party. He called for grange delegations to convene at Owatonna September 2, 1873, to effect a new political organization. His call alarmed state grange leaders. A month before the convention was to meet, Master George I. Parsons fired off letters to subordinate chapters, ordering them to ignore Donnelly's urgings. At the same time, he warned members of the consequences of following false prophets. "Prohibition [against partisan politics] is our only safeguard against sure and speedy destruction. Upon obedience to this law depends our very existence as an order." [6]

[3] Harrison Lowater to Donnelly, March 21, 1873, Charles Wood to Donnelly, April 5, 1873, Thomas T. Smith to Donnelly, April 5, 1873, William Paist to Donnelly, April 19, 1873, S. D. Hillman to Donnelly, May 15, 1873, Frank J. Mead to Donnelly, May 21, 1873, Andrew J. Murphy to Donnelly, May 26, 1873, Donnelly Diary, May 31, 1873, all in Donnelly Papers.

[4] Minneapolis *Evening Times and News,* June 18, 1873.

[5] J. F. Lewis to Donnelly, July 30, 1873, F. A. Elder to Donnelly, June 19, 1873, J. L. MacDonald to Donnelly, June 20, 1873, Isaac A. Christlieb to Donnelly, July 14, 1873, Dr. W. W. Mayo to Donnelly, July 21, 1873, Horace White to Donnelly, July 25, 1873, all in Donnelly Papers.

[6] Minneapolis *Farmers' Union,* August 16, 1873.

The opposing directives immediately divided grangers. Donnelly's flamboyant personality and his promise to return government to the people rallied many members to his side. One dissident explained his decision in an open letter to readers of the *Farmers' Union*. "We would say to Bro. P[arsons] that there is no danger of us (as he says) bartering away our beloved order for a mess of political pottage, but we are going to try to have a little sauce with our pottage, by electing good and true men to office, and give those political hucksters a chance to stay at home and learn the principles of honesty." In sharp contrast, conservatives in the order backed Parsons. Editor W. J. Abernethy of the *Farmers' Union* remained loyal to the state master. An editorial on August 16, 1873, listed his reasons for siding with Parsons. In the same article, Abernethy gave a stern warning to patrons not to follow Donnelly or anyone else advocating the development of independent political parties. "Given over to the hands of politicians, it [the Grange] will be manipulated for their own selfish interests and eventually [will be] split up under the leadership of various factions, and its usefulness entirely destrayed [*sic*]." [7]

Countering, Donnelly defended his actions in a pamphlet entitled "Facts for the Granges." The controversial Minnesotan noted how he alone had not been responsible for the third-party spirit; it had been prompted by dissatisfied farmers who were seeking relief from monopolies whose control over politicians of existing parties was stifling rural progress. In the tract, Donnelly admitted pulling the Grange into the thick of partisan politics and encouraging the formation of a third party. But according to Donnelly, veteran politicos had forced his hand. He said they had been "willing, perforce, that you should organize Granges, wear regalia, meet o' nights, discuss your wrongs and resolve to seek a remedy, but 'don't go into politics!' 'You will hurt some one.' Who? Why the fellows who are counseling you. . . . We are told the Granges are not political. True. But if the discussions in the Granges lead the members to certain political conclusions, they have a right to put those conclusions into force at the ballot box." [8]

On September 2, 1873, deputations from every section of Minnesota massed at Owatonna for a political convention. Sessions were kept open in the sense that nongrangers and grangers alike had been in-

[7] William Paist to Donnelly, August 18, 27, 1873, Christlieb to Donnelly, August 11, 18, 1873, O. F. Brand to Donnelly, August 5, 25, 1873, Joseph Goar to Donnelly, August 15, 1873, all in Donnelly Papers; Minneapolis *Farmers' Union*, August 16, 23, 30, 1873; Madison *Western Farmer*, August 23, 1873.

[8] Ignatius Donnelly, *Facts for the Granges* (n.p., 1873), 18–19.

vited to participate. Convention delegates heard speeches, drafted platform resolutions, and named candidates to oppose Republican and Democratic contenders. The new party was dubbed the Anti-Monopoly Party because of its pledge to emancipate farmers from the millstone of big business. Prior to election day, signs pointed to a successful debut for the new political organization. Its rallies were well attended, and people were discussing its possible impact on the destiny of Minnesota. With expectations running high, participants in the third-party movement looked ahead to election results. Hopes for a new day all but dissipated, however, after the votes had been tabulated. The state's electorate gave the new party a stunning defeat at the polls. Only five grangers won election to the state senate, and thirty-five others gained seats in the Minnesota House of Representatives. Donnelly was one of the winners; he was elected to the state senate.[9]

The poor election results did not discourage Donnelly and his associates; their positions hardened. They now decided to carry their fight to the next session of the state grange because they believed that such men as Parsons had betrayed the farmers' cause. After all, the conservatives had severed the order from the new party. Donnelly had become a celebrity in the order by the time the state grange convened. Upon entering the hall, he was "greeted with great applause & loud cries for a speech." Master Parsons tried to keep Donnelly from addressing the assembly, but according to the eager, politically minded granger the "audience grew clamorous and I came forward." From the reaction, Donnelly sensed that "the grangers are ripe for political revolution." Great applause at the meeting and subsequent receipt of many letters praising his efforts convinced Donnelly that there was growing sentiment for the Anti-Monopoly Party, so he continued fighting for the cause in 1874. And to keep his supporters abreast of developments, Donnelly began publishing a newspaper called the *Anti-Monopolist*. At the same time, he actively sought recruits to assist him. To this end, he sent grange secretaries circulars requesting them to furnish him with names of men in sympathy with his efforts.[10]

9 *Ibid.*, 19; Ridge, *Ignatius Donnelly*, 149–61; Ridge, "Ignatius Donnelly and the Granger Movement in Minnesota," 693–709; Madison *Western Farmer*, February 28, 1874; Chicago *Industrial Age*, January 24, 1874.

10 Donnelly Diary, December 17, 18, 1873, L. Jones to Donnelly, March 31, 1874, Daniel Tenny to Donnelly, May 17, 1874, D. F. McDermott to Donnelly, May 25, 1874, Christlieb to Donnelly, June 1, 1874, William Sleight to Donnelly, March 23, 30, August 6, 1874, all in Donnelly Papers; St. Paul *Anti-Monopolist*, 1874–75.

Although the Grange had not yet sanctioned his political activities, Donnelly believed approval would come. Knowing how essential an endorsement from the order was to the success of the Anti-Monopoly Party, Donnelly tried desperately to win over the opposition to his side. The politician-granger hoped to silence his critics by retreating somewhat from his original adamant position on the order and partisan politics.[11] An article written originally for the *Anti-Monopolist* and reprinted in the *Grange Advocate* contained Donnelly's revised and softened stand.

> The Grange has no right to talk politics as politics, no right to hold conventions or make nominations. But it has ... a perfect right to discuss questions of political economy ... transportation ... water communication ... patent laws ... [and] postal laws. ... These discussions should be conducted in an impersonal, unpartisan, and fraternal manner. If these discussions lead Granges to conclusions adverse to any political party or parties, they can put those conclusions into force outside of the Grange, and can as private citizens enter into any political movements they see fit.[12]

Since Master Parsons did not subscribe to even this view, Donnelly decided to campaign for his removal from office. Dissident grangers had been moving in that direction for months; with news of Donnelly's announced plans, radical members stampeded behind their leader and heeded his call for a campaign to oust Parsons. Bad feelings resulted, and the chasm between Parson's tradition-minded followers and rebels deepened and widened. Weeks before the state grange convened late in December, it appeared that the dispute would be the main order of business at the conclave. Donnelly won a major victory at the 1874 session of the state grange; Parsons was dismissed from the mastership. Donnelly was very pleased, and he gloated over the disposing of his adversary. His feelings were summed up in an editorial blast at the departed leader. "The State Grange draws a long breath of relief. The old man of the Mountain, who has so long ridden it, is no more. Parsons is dropped out of sight. ... He preserved the Republican party in power at the expense of the Patrons of Husbandry. He tried ... to reduce its members to silence and nothingness." [13]

In the end the victory at the 1874 session proved costly. Animosity engendered by the Parsons affair sparked a mass withdrawal from the

---

[11] St. Paul *Grange Advance*, July 22, 1874.   [12] *Ibid.*
[13] St. Paul *Anti-Monopolist*, December 17, 24, 1874.

organization, and its impact led directly to the termination of Donnelly's career as a granger-politician. A finale came in 1875 after the state house of representatives had blocked his bid for election to the United States Senate. Until this time, Donnelly had placed great faith in his grange membership and in his popularity among farmers, thinking that these would be sufficient credentials to give him a seat in Congress. Upon learning otherwise, he quit the agrarian brotherhood and joined the Greenback Party.[14]

The Minnesota Anti-Monopoly Party represented only one of many attempts to place the Grange more directly in politics; there were also determined efforts in Iowa, Illinois, and Wisconsin. In each case, dissidents broke with their state leaders and worked for the formation of third-party movements in their states. At no time, however, were these moves as carefully made as Donnelly's in Minnesota. Two factors accounted for differences. No one comparable to the fiery Minnesotan emerged in the other states to lead revolts, and the state grange leaders of Iowa, Illinois, and Wisconsin were stronger and more decisive than Master Parsons. Their quick and effective use of threats gave them enough power to keep chapters under control and to put down activists wanting to rebel.

In Wisconsin, the Dodge County Council of Granges issued a call for all members interested in forming a new party to gather at Milwaukee August 21, 1873. The invitation opened a spirited debate. Secretary James Brainerd of the state grange was the first of several state officers to react negatively to the announcement. He fired off an open letter to the Milwaukee *Sentinel* in which he warned members "to beware of the machinations of those who desire to thus violate one of the fundamental principles of the Order by casting us into the turmoil of political strife" and cautioned "all Granges against taking any notice whatever of the 'call' abovementioned, as it [is] entirely unauthorized by any proper authority." [15]

The same Milwaukee paper carried the Dodge County Council's reply August 7, 1873. Speaking for that body through an open letter to Brainerd, Secretary D. B. Bolens noted a modification of his organization's position. Instead of a meeting at Milwaukee, Dodge County grangers now proposed that "a convention of the Patrons of Hus-

14 Minnesota *House Journal* (1875), 53, 60, 66, 74, 81, 92, 127. For a complete recapitulation of Donnelly's up-and-down career as a reform politician, see Ridge, *Ignatius Donnelly*.
15 Madison *Western Farmer*, August 9, 1873; Milwaukee *Sentinel*, August 1, 1873.

bandry of Wisconsin be held at Watertown, on Tuesday, August 26th, 1873, for the purpose of consultation, and the transaction of business pertaining to the interests of the order, said convention to consist of the officers of the State Grange, state deputies, masters of subordinate granges, and such other members of the order who may wish to attend." [16] The Watertown meeting was held on schedule. Master John Cochrane and the executive committee represented the state grange. They and other delegates attending the convention met behind closed doors, but news still leaked out of the chamber. Dodge County's scheme for formation of a third party was rejected by the delegates. This development represented a major victory for the state grange over advocates of a granger political party.[17]

While Dodge County grangers were pushing for the establishment of a third party, patrons from Oregon, Rutland, and Brooklyn, Wisconsin, gathered at Brooklyn Grange No. 54 to endorse a reelection bid of Governor C. C. Washburn. In taking this action, they noted that they had been particularly impressed with "the able and impartial manner in which he has executed the trust reposed in him, in guarding the rights of the public against the encroachments of monopolies and unjust claims." Although an order rule against partisanship had been clearly violated by the resolution, no actions were taken against those grangers who voiced support for Washburn's candidacy.[18]

By their covert and overt activities, Wisconsin grangers engaged the state's professional politicians in a guessing game. Candidate endorsements, third-party possibilities, and widespread disenchantment with existing governmental policies created so much uncertainty in the state's political picture that veteran campaigners often lost their confidence and party chieftains no longer knew what to expect from rural voters. Each election produced anxious moments for party officials. GOP leader A. E. Bleekman was one of many who became nervous about election outcomes. His fears were revealed in a letter to Republican State Committee Chairman Elisha W. Keyes. Bleekman confessed to uncertainties and apprehensions about the 1874 elections. He was not sure whether the GOP was going to make a stronger showing in 1874 than it had in previous years because "it is impossi-

[16] Madison *Western Farmer*, August 16, 1873; Milwaukee *Sentinel*, August 7, 1873.
[17] Milwaukee *Sentinel*, August 8, 27, 1873.
[18] *Ibid.*, August 9, 1873; Madison *Western Farmer*, August 16, 1873.

ble to tell what the *grangers* will do." [19] As election day neared, Keyes became so alarmed that he ordered party worker D. W. Ball to survey the state's granges. Ball's compilation listed every grange in the state and noted political affiliations of every Wisconsin master and secretary. In submitting his listing, Ball advised Keyes to infiltrate granges with Republicans. "You will see from the marking[s that] they [the officers] are evenly divided Republican & Democrat. I am sory [*sic*] that so few Rep. join the Grangers. if [*sic*] a few more go in we can then hold the Old Ship Level. I advise every good Rep. to come in." Ball's intelligence report was of little use in the November elections because Keyes did not receive it until mid-October. The hour was too late for action; the GOP was left with no alternatives but to hope for the best.[20]

After three years of trying, activists in the Wisconsin State Grange finally scored a victory. Many members expressed disappointment in 1875 with results derived from the organization's neutral course, so they directed their representatives to the state grange to devise and initiate a more ambitious campaign for procurement of better public servants. Acting on this advice and on a resolution adopted by the state body, the chapter's secretary sent comprehensive questionnaires on railroad controls to every candidate for the state legislature. For each of thirteen questions pertaining to the operation and management of railroads, aspirants were asked to indicate whether they favored more or less governmental interference and regulation. Data obtained from this survey must have been helpful since the state grange used similar questionnaires later in the decade and in the 1880s.[21]

Meanwhile, Iowa grangers were contributing to the formation of an antimonopoly party in their state. Although the new political union had not been organized with order backing, it nevertheless was an illegitimate son of the First Granger Movement. By bringing more Iowa farmers together under one banner than any previous organization, the Grange had been able to coalesce farmers' ideas into a compact of grievances. It was greater awareness of agrarian problems, along with a joint resolve for action, that fostered the birth of a party

19 A. E. Bleekman to Elisha W. Keyes, July 15, 1874, in Elisha W. Keyes Papers, Wisconsin State Historical Society Library, Madison.

20 D. W. Ball to Keyes, October–November, 1874, in Keyes Papers.

21 Wisconsin State Grange, *Proceedings*, 1875, Appendix, 1–2; 1877, pp. 60–61; 1881, pp. 8–10.

devoted to granger programs and principles. Stated briefly, the agrarian fraternity served as a catalyst for third-party activities. It put momentum in the farmers' movement for a better life, but it did not consciously drive its members to a political course.[22]

Many grangers actively backed the Anti-Monopoly Party, but their actions were not sanctioned by state grange spokesmen. Renouncements from them, however, did not deter members bent on involving the order in third-party affairs. In fact, their positions often solidified; the twelfth article of the grange constitution barring partisan politics for many dissidents was no impediment because they placed the order's programs ahead of its rules. In their opinion, violations of this article were noble because they afforded members an opportunity to work directly for organization objectives; obeisance curtailed grangers and killed the order's chances for success. Although the activists' arguments had some merit, no move was made by the state grange to accommodate grangers wishing to tie the order directly to the Anti-Monopoly Party. The order in Iowa came closest to involvement on June 7, 1873, when officials William D. Wilson and M. L. Devin attended a convention of the party. After a day of discussions with third-party leaders, Wilson and Devin chose not to commit their organization to the party's activities.[23]

As thoughts of grange involvement in the Anti-Monopoly Party were fading, a new consideration emerged. On March 31, 1873, grangers from twenty Iowa counties converged at Waterloo to nominate Dudley W. Adams and James Wilkinson for governor and lieutenant governor respectively. These drafts were attempted because both men possessed attractive credentials for public office. Adams and Wilkinson had already demonstrated leadership skills while serving as officers in the state grange, and both men had displayed keen knowledge of agrarian problems and great compassion for rural ideals. Although Adams had been flattered by the trust shown him, he still removed himself from contention. His decision not to run had been swayed by the order's prohibition against partisan politics. Adams was convinced that it would be "most injudicious to divert it [the Grange] from its original plan" because a move by him to participate actively

22 Mildred Throne, "The Anti-Monopoly Party in Iowa, 1873–1874," *Iowa Journal of History*, LII (October, 1954), 293–94; Waukon, Iowa, *Standard*, October 2, 1873; Des Moines *Iowa Homestead and Western Farm Journal*, August 2, 1872.

23 Waukon, Iowa, *Standard*, October 2, 1873; Throne, "The Anti-Monopoly Party in Iowa," 291. Des Moines *Iowa Homestead and Western Farm Journal*, April 18, 1873, December 6, 20, 27, 1872.

in politics would tend "not only to defeat the very object aimed at in the present, but ... [it also would endanger] our usefulness in the future." [24]

Contemporary with political stirrings in Minnesota and Iowa was a massive drive to commit the Illinois State Grange to the Farmers' and People's Anti-Monopoly Party. This effort was primarily the handiwork of S. M. Smith and W. C. Flagg, secretary and president respectively of a loose federation of independent farmers' clubs known as the Illinois State Farmers' Association. Smith belonged to the Grange as well as the independent organization, but Flagg limited his activities to the farmers' association. He refused to join the Grange because it had certain features which he found objectionable. Nor did Flagg hesitate to criticize the order's weaknesses. "Its faults ... are its secret organization and its deprecation of political action in its combined capacity." [25]

Meanwhile, the Grange was throwing wrenches into the political machinery of at least seven other states. Briefly in 1873 it looked as though Hoosier grangers were going to upset traditionally Republican Indiana by running an independent slate of candidates the next year, but the threat ebbed as the election neared. In fact, Republican prospects had improved so much by January 11, 1874, that party aide T. H. Bringhurst gave assurances to State GOP Boss Daniel D. Pratt:

It may safely be assumed that there will be no Granger ticket for the State.... The Granger effort will be directed chiefly to county and local tickets. They will be generally either granger county tickets or independent tickets supported by Grangers.... They [the farmers] will be so circumstanced by the manner of their election: the professions made by them and their friends in the camp[aign] and by the general desire for reform expressed by the movement as its chief basis, that, if care is taken now, the Grange vote may, without any sacrifice, be secured. [26]

As early as 1874 and 1875, Master William W. Lang of the Texas State Grange was brought forward as a possible gubernatorial candidate in the Lone Star State, but he disclaimed all efforts in his behalf. In the latter year, a campaign was also launched to put his name in nomination for Congress from the fourth district; and again a boom was halted at Lang's request. Before the spring elections of 1876, he was mentioned again for public life, this time for a seat in the state

---

24 Des Moines *Iowa Homestead and Western Farm Journal*, April 11, 25, 1873.
25 Chicago *Tribune*, August 9, 1873; Chicago *Industrial Age*, August 30, 1873.
26 T. H. Bringhurst to Daniel D. Pratt, January 11, 1874, in Daniel D. Pratt Papers, Indiana State Library, Indianapolis.

legislature. He accepted this draft. His bid for the legislature proved successful, but rule-bound grangers from Texas and elsewhere marred the victory with accusations. He was charged with dragging the Grange into partisan politics.[27]

From 1876 to 1878, Lang's name was considered repeatedly by the agrarian press as the farmers' choice for governor. To these efforts, the master replied that his political aspirations were dependent upon the desires of the Democratic State Convention. When the party finally converged at Austin in July, 1878, it did not bring forward the granger as its candidate. Firm to his pledge not to seek the governorship unless the Democrats drafted him, Lang refused to make the race as an independent. He soon dropped out of the picture, and the mastership passed to A. J. Rose, a man committed to political neutrality.[28]

Dakota Territory grangers also dabbled briefly in partisan politics. When antimonopolists met in plenary session at Elk Point June 13, 1874, patrons were conspicuously represented. As grangers were preparing a statement in behalf of independent politicians, Territorial Master E. B. Crews was busy disavowing their actions. His decision not to allow the Grange to be identified with the reformers' efforts no doubt was influenced by personal convictions. Crews had long been active in GOP affairs. This master controlled a sizable portion of his following, not as a Republican or as an independent, but as a granger. Order opponents emphasized Crews's dual position, and their attacks served to discredit both the Grange and its leader. Thus, whatever the private motives of Master Crews, his actions compromised the Grange and contributed to its untimely decline.[29]

For the first few years of its existence, the Missouri State Grange avoided partisanship. In its place, the state chapter sought to "inculcate a spirit of independence in politics which would lead the members to vote for their own interests." Members who wanted to select and endorse candidates at grange meetings disliked this policy from the beginning and fought for its repeal. Dissidents often bent the rule against political action to their advantage. Technically, grange proceedings began with a convocation of order rules and ended with a suspension of rules. Therefore, if rules were suspended, grangers would be theoretically free to discuss anything and endorse anyone.

[27] Roscoe C. Martin, "The Grange as a Political Factor in Texas," *Southwestern Political and Social Science Quarterly*, VI (March, 1926), 369–70.
[28] *Ibid.*, 370–71.
[29] Schell, "The Grange and the Credit Problem in Dakota Territory," 77–78.

Working on this assumption, many granges in Missouri and elsewhere sidestepped the constitution to proceed with discussions on partisan issues. Farmer-politician David Rice Atchison of Missouri supported these efforts. He felt "the people have the remedy [for solving political problems] in their own hands, and if they do not use it, they alone are to blame, and if they fail it proves the incapacity of the people for self-government." [30] Eventually, rebels gained control of the Missouri State Grange, and as a result of their successful insurgency, several Missouri grangers sought public offices during the nineteenth century on the strength of their membership in the order. Former master Henry Eshbaugh of the state grange ran for Congress on the Greenback ticket in 1878, and Executive Committeeman Ahira Manring and Granger G. B. Debenardi were the Union Labor Party candidates for governor and register of lands, respectively, in 1888. Master D. N. Thompson of the state grange engineered the Union Labor Party campaigns the same year.[31]

Thompson's questionable activities did not set a precedent. There had been similar involvements earlier in the century. During the late 1870s, Gus W. Richardson doubled as secretary of the Kentucky State Grange and president of the Meadville Greenback Club. On May 30, 1874, the Raleigh *News* disclosed the candidacy of a state grange official. E. C. Davidson declared himself a farmers' candidate for Congress from the sixth district of North Carolina. His declaration provoked Editor Richard T. Fulghum of the *State Agricultural Journal*, who was distressed because the aspirant had been serving as steward of the state grange prior to his announcement. When Davidson commenced his campaign, he made it clear that he was giving up his post in the state grange. Still, this concession did little to placate Fulghum. This newspaperman was so absorbed with the idea that the order rule against political partisanship would be threatened as long as Davidson remained in the organization, that he insisted on the candidate's complete withdrawal from the order.[32]

In compliance with order rules, some granges were quick to denounce incompetent public officials but were hesitant to name sub-

---

30 Homer Clevenger, "Agrarian Politics in Missouri, 1880–1896" (Ph.D. dissertation, University of Missouri, 1940), 70–71; David Rice Atchison to Judge D. H. Birch, Sr. August 24, 1874, in David Rice Atchison Collection, Western Historical Manuscripts Collection, University of Missouri Library, Columbia.

31 Clevenger, "Agrarian Politics in Missouri," 71–72.

32 Louisville *Farmers Home Journal*, October 31, 1878; Raleigh (N.C.) *State Agricultural Journal*, June 4, 1874.

stitutes. A good case in point occurred early in 1874. Ceres Grange No. 1 of Colorado challenged the prodigal legislative records of state representatives Levi Harsh and D. H. Nichols. The two men eventually were censured by the subordinate grange for supporting a transfer of the capital from Denver to Pueblo and for backing the formation of three state colleges. The chapter concluded that Harsh and Nichols were not worthy to represent the people of Colorado in the state legislature because they were too extravagant with tax revenues. Although members of Ceres Grange had called for the two legislators' resignations, no subsequent measures were taken to suggest replacements.[33]

A more notable example of a public servant being rebuked by grangers involved a member of Grover Cleveland's second administration. Secretary of Agriculture J. Sterling Morton incurred the hostility of the Grange by stating in an address to the 1893 session of the Farmers' Auxiliary Congress that "the most insidious and destructive foe to the farmers is the professional farmer, the promoter of granges and alliances. . . ." [34] Morton's disparaging reference to the men and women who were working for the farmers' brotherhood was taken as an insult by grangers. They adjudged from the remark that Morton was not fit to represent farmers in the president's cabinet. His statement showed contempt for rural leaders, and it demonstrated a lack of understanding of agrarian organizations. Since granger displeasure with the secretary's conduct was not considered a partisan issue, no effort was made to curtail campaigns aimed at forcing Morton to resign from his post. As a result, local after local and the National Grange denounced the secretary of agriculture and pushed for his ouster.[35]

In 1896, to avoid a repetition of the Morton affair, grangers petitioned President-elect William McKinley to appoint a secretary of agriculture who sympathized with the farmers' cause. Several granges offered McKinley the name of Colonel J. H. Brigham, master of the National Grange, as the man best suited for the position. Although

[33] Denver *Rocky Mountain Weekly News*, February 11, 1874.

[34] Rhode Island State Grange, *Proceedings*, 1894, p. 27; National Grange, *Proceedings*, 1893, pp. 27–28, 86–87, 161–62.

[35] National Grange, *Proceedings*, 1893, pp. 27–28, 86–87, 161–62; Ohio State Grange, *Proceedings*, 1893, pp. 8, 16–17; New Hampshire State Grange, *Proceedings*, 1893, pp. 17–18; Illinois State Grange, *Proceedings*, 1893, p. 34; E. C. Hutchison to Henry William Blair, December 30, 1893, Henry William Blair Collection, in Western Historical Manuscripts Collection, University of Missouri Library, Columbia.

the grange official was not chosen, Patrons of Husbandry still were pleased with the chief executive's choice of James Wilson.[36]

In some parts of the country, the Farmers' Alliance posed a serious threat to the Grange's ban on partisanship. Alliance rules permitted political endorsements and nominations. The power given the Texas Alliance to piece together a political organization precipitated a feud within the state grange hierarchy in 1886. Secretary R. T. Kennedy believed the state grange would have to follow the lead of the Alliance and establish its own party if it wanted to compete with its rival for members; Master A. J. Rose disagreed. He alleged that such a course would be unnecessary and contended that formation of a political organization would compromise the order's position as a neutral force in politics. Two divergent opinions quickly led to a division in the ranks. One side remained loyal to Rose, while the other faction backed the activist urgings of Kennedy. The latter group often defied organization rules by taking up politics. After five years of internal bickering, the state grange decided a change in leadership would benefit the order. As a result, Rose was asked to step aside for John B. Long, congressman, overseer of the state grange, and recipient of the mastership.[37]

After many early attempts, the National Grange's determination to remain impartial in political matters was tested again in 1892 when the Populist Party ran candidates on a platform dedicated to agrarian objectives. But once again the order remained true to its constitution. As a result of the order's neutrality and other factors, the new political organization had no apparent impact on granger voting habits. For example, in twenty New York counties where grange membership exceeded 500, not one went for James B. Weaver, the People's Party candidate for president. He ran a poor fourth behind the Republican, Democratic, and Progressive party nominees in nineteen counties and also trailed these three and the Socialist Labor Party candidate in the twentieth county. Election results showed Grover Cleveland winning in Seneca County and Benjamin Harrison carrying all the others.[38]

36 Wisconsin State Grange, *Proceedings*, 1896, p. 39; Massachusetts State Grange, *Proceedings*, 1896, pp. 62–67; Illinois State Grange, *Proceedings*, 1896, p. 56; National Grange, *Proceedings*, 1896, pp. 111–12; Maine State Grange, *Proceedings*, 1897, p. 57.

37 New York State Grange, *Proceedings*, 1890, pp. 54–56; 1891, pp. 129–30; Illinois State Grange, *Proceedings*, 1889, p. 45.

38 Smith, "The Grange Movement in Texas," 304–306; New York State Grange, *Proceedings*, 1893, p. 122; *The World Almanac*, 1893 (New York, 1893), 355.

Four years later the Democratic Party stole the thunder from the Populist Party platform and put forth a candidate friendly to agrarian interests. But the Grange still remained impartial. William McKinley was definitely more at odds with granger philosophy than William Jennings Bryan, yet no favoritism was shown. This might account for the strength of the Republican nominee in New York; McKinley won easily in every bastion of grange strength.[39]

Throughout the late nineteenth century, then, many prominent politicians belonged to the Grange, and no one challenged their presence. In fact, their memberships were often sources of pride. These dignitaries were regularly called upon to deliver speeches at state and national granges, and notations were made about them in agricultural journals. The list of famous grangers included Governors William R. Taylor of Wisconsin and D. H. Goodell of New Hampshire; Lieutenant Governor E. F. Jones of New York; and Speaker of the Kansas House of Representatives Samuel Newitt Wood.[40]

In 1892 Granger Mortimer Whitehead summarized the role of his order in politics for *The American Journal of Politics.* "Its teachings are full of pure politics, but is *partisan* NEVER. It cannot be used as a cat's paw to draw political chestnuts out of the fire either for parties or for individuals. . . . The Grange is not a good place for one who loves his party more than his country; or for one who makes the organization or his party secondary to his own personal ambition. It cannot be used by such as a stepping-stone to political preferment." [41] Although there had been many exceptions to the summary drawn by Whitehead, his conclusion still depicted the true aspirations of the brotherhood in the nineteenth century. The Grange honestly did try to sever partisanship from its programs, and it did denounce almost every dissident move aimed at forwarding either parties or would-be office seekers.

39 *The World Almanac,* 1897, p. 453; New York State Grange, *Proceedings,* 1896, pp. 54–55.

40 Kansas *House Journal* (1877), 3–5; Madison *Western Farmer,* November 15, 1873; Charleston *Rural Carolinian,* December, 1873; New Hampshire State Grange, *Proceedings,* 1890, pp. 48–49; Richmond *Virginia Patron,* May 4, 1877.

41 Whitehead, "The Grange in Politics," 115.

CHAPTER IX

The Grange and Miscellaneous Legislative Programs

Grange leaders wanted no part in direct involvement in partisan politics. Instead, they sought to channel agrarian discontent into nonpartisan drives for legislative solutions to political problems. To this end between 1870 and 1900 the order unfolded a complex series of demands. These fell into ten major categories. Included were requests for structural changes in government, temperance laws, democratic reforms, immigration restrictions, conservation measures, revisions in the bases of money, alterations in taxes and tariffs, protection for consumers, educational improvements, and tighter control of transportation.[1]

Perhaps surprisingly, conservation emerged as a primary concern of grangers. The public domain was disappearing rapidly and wildlife was vanishing at an alarming pace. An urgent need was developing for governmental programs to arrest the ruthless and indiscriminate exploitation of the nation's resources. Accordingly, grangers responded with demands for such legislative curbs to check the trends affecting adversely their interests as farmers.

One of the first matters to claim grangers' consideration in the general area of conservation was the rapid disappearance of public land. In their opinion, the public domain should be set aside specifically for bonafide settlers; it was not intended to be the bailiwick of "capitalists, corporations, and speculators." To insure the letter of the law, according to the Grange, the government should take up every acre held for investment purposes and hold it for actual settlers. In making their pleas for agricultural settlement of the public domain, grangers conveniently ignored the fact that they as farmers represented one bloc of land speculators in the country. Nevertheless, with all of their inconsistencies, grangers still deserve some credit for focusing at-

---

[1] Transportation will be discussed in detail in the following chapter. See earlier chapters for granger legislative demands pertaining to education.

tention on the pressing need to preserve for future generations what was left of a rapidly fading frontier.[2]

When the full ramifications of the settlement issue were understood, the resolve to open land exclusively to agricultural use was not shared by everyone in the order. In fact, the question of what the government's attitude should be in regard to western lands led to a division in the order's ranks. Regional considerations contributed greatly to the split. Eastern members had nothing to gain and much to lose if additional land were released for settlement since every new acre placed under cultivation meant lower land values and greater competition for them. On the other hand, western grangers needed increased acreage to make their farming operations successful. Arid conditions in much of the West necessitated larger farm units. This conflict of interests came out into the open when western grangers urged the National Grange to adopt resolutions favoring government-sponsored irrigation projects. Easterners united to thwart these efforts. The end result was a compromise which failed to placate either group. In 1892 the National Grange rejected a proposal to campaign for water projects at federal expense, but it left the door open for states to undertake irrigation programs.[3]

In contrast to questions of irrigation and land development, the subject of wildlife protection posed few problems for grangers. There was a general agreement among members regarding the need for effective game laws to save all but predators like coyotes and wolves. Grangers were particularly concerned about the welfare of insect-eating birds. Their role in the control of certain pests was well known to farmers. For this reason, grangers wanted to guard insectivores from "the lawless, careless hunters from the cities and villages." Granges throughout the country regularly petitioned legislatures to enact laws to protect quail and prairie chickens from sportsmen's guns.[4]

2 Michigan State Grange, *Proceedings*, December, 1875, p. 96; Fite, *The Farmers' Frontier*, 19; Denver *Colorado Farmer*, January 22, 1885; National Grange, *Proceedings*, 1886, p. 73; 1888, pp. 116–17.

3 Massachusetts State Grange, *Proceedings*, 1889, p. 25; 1890, pp. 68–69; New Hampshire State Grange, *Proceedings*, 1891, p. 151; Denver *Colorado Farmer*, January 22, 1885; California State Grange, *Proceedings*, 1886, p. 55; National Grange, *Proceedings*, 1892, p. 207.

4 Illinois State Grange, *Proceedings*, 1899, p. 39; 1890, p. 73; Pennsylvania State Grange, *Proceedings*, 1887, 37; Charleston *Rural Carolinian*, December, 1874; Topeka *Kansas Farmer*, November 11, 1874; Indiana State Grange, *Proceedings*, 1882, p. 51; National Grange, *Proceedings*, 1878, p. 67; Alabama State Grange, *Proceedings*, 1874, p. 25.

A number of other conservation measures also received order en-dorsement. Pennsylvania members backed their state forestry associa-tion's attempts "to prevent the complete denudation of our valleys and hillsides." Meanwhile, New Hampshire grangers asked their state legislature to consider the feasibility of restocking rivers and streams with salmon and shad.[5]

At the local level, grangers set examples in the field of conservation for others to follow. On the prairies, grange-sponsored arbor days were popular. For example, Providence Grange of Kansas set aside April 17 as its arbor day. Members competed to see who could plant the most trees. Winners were exempted from paying membership dues for two quarters.[6]

Matched to the order's pleas on behalf of conservation were its de-mands for laws to prohibit inhumane treatment of animals and to protect consumers from products of diseased livestock. Massachusetts grangers were latecomers to these battles, but they did add their voices to the movement for whatever value this had. To reduce maladies among dairy cattle, the state grange went on record in 1898 in favor of legislation to give the Massachusetts Dairy Bureau authority to in-spect barns and dairies and to grade the buildings. Four years earlier the same body had proposed a state-sponsored incentive program aimed at coaxing farmers to rid their herds of tuberculous animals. Under the program, the state would pay "the owner thereof, out of the treasury of the commonwealth, if such animal has been within the State six months continuously prior to its being killed, provided such person shall not have, prior thereto, willfully concealed the existence of tuberculous, or by act or willful neglect contributed to the spread of such disease." Other granges asked for appropriations to reduce the inroads of hog cholera and gypsy moths and to place bounties on chicken hawks. Patrons in New York asked their state assembly to "in-vestigate the matter of dehorning cattle and report to the Legislature their findings, that a law may be enacted defining whather [sic] it is not an inhuman practice, but one which does away with the liabil-ity to danger by goring, which is always present in any herd of horned cattle, and adds greatly to the comfort of such herd and to the profit

5 Pennsylvania State Grange, *Proceedings*, 1886, pp. 60–61. For a similar plea, see Alabama State Grange, *Proceedings*, 1874, p. 25; New Hampshire State Grange, *Proceedings*, 1874, p. 36.
6 Topeka *Kansas Farmer*, June 30, 1875.

of the owner." A bill permitting the removal of horns became law, thus giving the state grange a victory.[7]

Granges in communities without regulations against stray dogs often sought appropriate legislation to deal with menacing canines. Protective safeguards against strays had long been recognized as necessary by local governments, and their councils passed a variety of municipal ordinances covering the problem. These, however, did not directly help farmers outside the reach of city laws. Additional action had often become necessary in many rural localities because packs of vicious dogs were ravaging flocks of sheep, and the individual farmer's only protection was his hunting gun. Knowing from experience the inadequacy of farmers, with rifles in hand, stalking and searching fields for troublesome dogs, granges looked for better solutions. Their examinations and discussions usually led to the same conclusion. Their hopes for alleviating the problem would have to be pinned to county dog-tax proposals. The philosophy behind the suggestion was simple. If county licenses were imposed, dog owners would likely keep their pets corralled because they now would represent an investment.[8]

A paradox developed in regard to stock protection. Many granges fought fence laws requiring enclosure of farm animals. Virginia and Pennsylvania chapters were especially adamant on the subject. Those states already had such requirements on the statute books when the order turned its attention to the matter of repealing them. Obviously disappointed with the costs involved in erecting pens and with the results, dissident farmers working through granges sought repeal of fence laws. In the end, these battles bore no fruit; lawmakers refused to listen to their appeals. Elsewhere, granges were known to support fence laws. Elgin Grange No. 75 of Wabash County, Minnesota, for one, pleaded for the enactment of such a law. Members of the chapter felt that cattle owners should be compelled to prevent their stock from running through neighbors' fields, and the only way to accom-

[7] Massachusetts State Grange, *Proceedings*, 1898, pp. 15–16, 101; 1894, pp. 30–31, 75; West Virginia State Grange, *Proceedings*, 1895, p. 22; Missouri *House Journal* (1877), 477; Pennsylvania *House Journal* (1891), 120; New York State Grange, *Proceedings*, 1893, pp. 90–91.

[8] Massachusetts State Grange, *Proceedings*, 1885, p. 70; 1888, pp. 19–20, 122; Raleigh, North Carolina, *State Agricultural Journal*, April 23, 1874; Virginia *House Journal* (1874), pp. 85, 134, 145, 168; (1876), 446; (1878), 424; Alabama State Grange, *Proceedings*, 1874, p. 25; West Virginia State Grange, *Proceedings*, 1895, p. 22.

plish this was by making it mandatory to keep stock behind barriers.[9]

Apart from conservation matters, grangers also battled for class legislation beneficial to agriculture. For one thing, they protested the manufacturing of vinegar from whiskey. In 1879 Congress sparked the controversy by approving the Whiskey-Vinegar Bill. Apple growers were the first interest group to complain. Making vinegar from distilled spirits especially hurt them because this process was much cheaper than waiting for apple cider to ferment. Soon, the National Grange and the Massachusetts State Grange were showing sympathy for the orchard owners' plight. Both groups waged vigorous but unsuccessful fights for repeal of the act.[10]

Sericulture was another concern of grangers. To the satisfaction of the National Grange and several subordinate chapters, silk culture would prove itself practical in many western and southern states. For this reason, grangers requested Congress in 1888 to approve the Thompson-Chandler Bill for the development and encouragement of silk culture in the United States under the supervision of the Commissioner of Agriculture. This push for sericulture proved to be a major fiasco because silk worms never thrived in the New World.[11]

Quite often in their drives for legislation, grangers asked for extensions of the arm of government to give consumers protection against poor quality products. Among other things, patrons wanted boot and shoe manufacturers to stamp on all footwear the product composition, bakers to post weights on loaves of bread, "growers and dealers in seeds . . . to state upon each package offered for sale a guarantee of the per cent of purity and germinating power contained in the same," and fertilizer producers to print content analyses on their products.[12] Granger reasoning behind these requests was sound. Chapters asked for passage of legislation giving these consumer safeguards because without them purchasers had no way of judging a product's quality

9 Pennsylvania *House Journal* (1895), 941, 1471; Virginia *Senate Jou nal* (1875), 112; Atlantic Council, Patrons of Husbandry, *Proceedings*, 1873, n.p.; Minnesota *House Journal* (1874), 72, 212.

10 *Congressional Record*, 46th Cong., 1st Sess., 1925; National Grange, *Proceedings*, 1882, pp. 97–98; Massachusetts State Grange, *Proceedings*, 1882, pp. 6–7.

11 *Congressional Record*, 50th Cong., 1st Sess., 5087; Massachusetts State Grange, *Proceedings*, 1888, p. 117; Bidwell and Falconer, *History and Agriculture in the Northern United States*, 101.

12 National Grange, *Proceedings*, 1891, p. 180; Colorado *House Journal* (1895), 202; Massachusetts State Grange, *Proceedings*, 1890, p. 80; Virginia *House Journal* (1877), 752; New York State Grange, *Proceedings*, 1895, p. 148.

except through trial and error. Since this procedure clearly gave the seller advantages over the buyer, grangers felt the government should act to give the purchaser a fair chance.

The introduction of oleomargarine into the United States from Europe in 1874 soon challenged grangers dependent on dairying for their livelihood. Since the new product could be processed in such a way as to make it indistinguishable from butter, it soon was falsely labeled as the more expensive spread. Had this fraudulent practice been allowed to continue, consumers would have been cheated and dairy farmers would have been ruined. Pennsylvania grangers early recognized the menace that the substitute represented to their interests. They therefore worked hard to keep it off the nation's markets. Their campaign in large part led to passage of the Pennsylvania Oleomargarine Law of 1878, one of the first statutes of its kind enacted by a state general assembly. This law required all oleomargarine producers to stamp their product as such, and penalty for failing to comply was a $100 fine. This initial act had many loopholes, so amendments were approved in 1883 and 1885 which banned all sales of the dairy substitute. Makers of the product pushed vigorously for repeal in the 1890s, but grangers matched their efforts with a barrage of petitions against legalization. The legislature stood firm and refused to annul the oleo law.[13]

The drive begun by the Pennsylvania State Grange against oleomargarine soon carried into almost every state where the order had active chapters and members engaged in dairying. For the most part, granger arguments followed the same line of reasoning. Legislation must be passed, said the grangers, because "butterine" defrauded consumers and crippled dairymen. At times, patrons also injected health considerations into their appeals. Members in New York alleged that the nondairy product was a carrier of "deadly germs of disease, such as trichinosis, tapeworm, glandeur, pleuropneumonia, and typhus." For this reason and this reason alone, the state grange said it was pledging to stamp out "this vile compound and death dealing 'oleo' " manufactured and distributed by "soul-less corporations." In stating their case, these New Yorkers noted that their claims had been supported by fiindings of a reputable scientist, Dr. R. E. Clark. A chemist by profession, Clark had charged before the New York Dairy and Food Commission that butter substitutes were unfit for human con-

13 Brenckman, *History of the Pennsylvania Grange*, 205–206; Pennsylvania *House Journal* (1895), 758–1034, *passim*; (1897), 2293, 2561–62.

sumption.[14] By 1900 many state legislatures had already prohibited selling oleomargarine as butter. One of the most comprehensive acts covering this subject was passed by the West Virginia General Assembly in 1891. This law declared it "unlawful for any manufacturer or vendor of oleomargarine, artificial or adulterated butter, to manufacture or offer for sale within the limits of this state, any oleomargarine, artificial or adulterated butter, whether the same be manufactured within or without the state, unless the same shall be colored pink." Speaking before the 1891 session of the state grange, Master Thomas C. Atkeson claimed the order had been directly responsible for the enactment.[15]

Correlating the Grange's demands for state legislation was the order's drive for federal action. The National Grange along with its subordinate and state chapters petitioned senators and representatives in the 1880s for a legal prohibition against misleading consumers with artificial spreads.[16] The campaign succeeded; Congress approved an oleomargarine bill which was subsequently signed into law by Grover Cleveland in 1886. After affixing his name to the bill, the chief executive explained what he expected the law to accomplish:

> If this article has the merits which its friends claim for it and if the people of the land with full knowledge of its real character desire to purchase and use it, the taxes enacted by this bill will provide a fair profit to both manufacturer and seller. If the existence of the commodity taxed and the profit of its manufacture and sale depend upon disposing of it to the people for something else which it deceitfully imitates, the entire enterprise is a fraud, and not an industry. And if it cannot endure the exhibition of its real character, which will be afforded by inspection, supervision and stamping which this bill directs the sooner it is destroyed the better for the interests of fair dealing; and, these same principles apply to the coloring of "Oleo" at this time.[17]

Pure-food laws were also high on the grange list of legislative demands. With no controls, all types of misbranded and adulterated foods, liquors, and drugs were being sold to an unsuspecting public.

14 New York State Grange, *Proceedings*, 1892, pp. 116–17; 1893, p. 100; New Hampshire State Grange, *Proceedings*, 1894, p. 19; Wisconsin State Grange, *Proceedings*, 1879, p. 47; New York State Grange, *Proceedings*, 1893, p. 100; West Virginia *House of Delegates* (1891), 269–70,
15 West Virginia State Grange, *Proceedings*, 1891, pp. 12–13.
16 *Congressional Record*, 49th Cong., 1st Sess., 4657; 50th Cong., 1st Sess., 253, 400, 1131; National Grange, *Proceedings*, 1888, pp. 117–18.
17 Massachusetts State Grange, *Proceedings*, 1888, p. 18; Allan Nevins, *Grover Cleveland: A Study in Courage* (New York, 1944), 362.

Yet governments were doing nothing to check these unfair practices, so granges began appealing for action. To secure positive legislation in this regard, chapters besieged Congress and state legislatures with petitions arguing the case for consumer protection. Although these particular drives appeared in the late 1870s, no comprehensive act was passed until Theodore Roosevelt's presidency. Then it was not so much the Grange that was responsible; a muckraker named Upton Sinclair did more to move the nation's lawmakers and chief executive in the direction of pure-food legislation.[18]

Often in seeking legislation, grangers rejected compromise because to them their demands represented questions of morality. Their battles against "evil" forces in society thus took on the characteristics of a negatively inspired evangelist admonishing his flock not to sin because damnation was the price of disobedience. Fear of consequences also filled many grangers with zeal and determination to rid society of vices.

There was a special resolve on the part of grangers to stamp out sales of intoxicants because many order members believed that prohibition provided the way to a better society. "Intemperance is the fountain from which issues many of the evils that affect society, and many of the corrupting influences that threaten the peace, prosperity, and perpetuity of our Government." Since the issue of prohibition had so many moral overtones, the Grange approached with unusual earnestness the problem of eliminating the liquor traffic. Legislators were asked countless times to enact temperance laws, grangers were requested to consider the facts "before casting their ballots for any one for office, Town, County, State or National who is in sympathy with the liquor traffic, or who is addicted to the habitual or even moderate use of intoxicating drinks," and chapters were ordered to purge their memberships of "habitual dram drinkers and dram sellers." By resolution, members also threatened to boycott agricultural fairs unless alcoholic beverages were kept off the grounds. In 1889 Illinois grangers even suggested "that the State Grange in the future establish its headquarters at hotels where no bars are kept." [19]

18 Texas *House Journal* (1879), 173; National Grange, *Proceedings*, 1888, pp. 117–18; 1896, pp. 74–81; Indiana State Grange, *Proceedings*, 1890, p. 30; *Pure Food*, Circular from the National Grange, dated August 13, 1888, in Clemson University Patrons of Husbandry Papers; George E. Mowry, *The Era of Theodore Roosevelt and the Birth of Modern America, 1900–1912* (New York, 1958), 207.

19 National Grange, *Proceedings*, 1878, pp. 105–106; New Jersey State Grange, *Proceedings*, 1891, pp. 55–56; Wisconsin *Assembly Journal* (1874), 140; Denver *Colo-*

Not everyone in the order appreciated the Grange's official position on alcoholic beverages. Friends of prohibition in the organization had three notable confrontations with adversaries of restriction. At the 1876 session of the Ohio State Grange, temperance-minded members demanded a strong stand against intoxicants and wanted every known liquor dealer suspended from the organization. Their motions were overruled. Two years later when the National Grange took up the question of alcohol, eleven members opposed the prohibitionist position being proposed to the parent body. Ironically, seven of the eleven negative votes were from the South, a region not known for its support of legal alcohol. A third confrontation occurred in 1891, when the general body of the New Jersey State Grange ignored the advice of the committee on the good of the order and adopted a prohibition resolution offered by Crosswicks Grange. It pledged "that we as an organization will do all in our power as law-abiding citizens to blot out this plague spot in the nation." Judging by the evidence, one must have to disagree with the high-ranking Illinois public official who reportedly said that "these d—d grangers are all prohibitionists." [20]

There were also moralists in the Grange who voiced strong opposition to the use of tobacco. The "pernicious weed," as they often called it, was condemned because it was considered injurious to the body and expensive as well. To discourage addiction to the "filthy habit," grangers pleaded with lawmakers to prohibit sales of tobacco products to minors and asked for educational campaigns to acquaint the public with the dangers of smoking. Of course, these demands did not take into consideration the welfare of tobacco growers who belonged to the order. Since their vested interests were at stake, they had something entirely different in mind when they appealed for legislation. What they sought for their product was protection, not its elimination. Specifically, they worked for fulfillment of three objectives. First, they wanted every state and federal tobacco tax repealed. Second, they asked the federal government to induce foreign nations to

rado Farmer, September 25, 1884; Indiana State Grange, *Proceedings*, 1879, p. 38; 1881, pp. 28–29; Social Grange, Secretary's Minutes, December 6, 1891; Michigan *House Journal* (1874), 95; Colorado State Grange, *Proceedings*, 1888, p. 18; Missouri State Grange, *Proceedings*, 1892, p. 11; Massachusetts State Grange, *Proceedings*, 1882, p. 41; Michigan State Grange, *Proceedings*, 1891, pp. 31–32; Indianapolis *Indiana Farmer*, October 5, 1878; Illinois State Grange, *Proceedings*, 1889, pp. 54–55.

20 Ohio State Grange, *Proceedings*, 1876, p. 50; National Grange, *Proceedings*, 1878, pp. 105–106; New Jersey State Grange, *Proceedings*, 1891, pp. 55–56; Indianapolis *Indiana Farmer*, March 31, 1883.

cut their duties on American tobacco. Third, they desired laws which would empower state governments to inspect and grade tobacco samples. Grangers questioned why "mere ownership of a warehouse should carry with it the privilege of weighing or sampling the produce which may be sold there." [21]

In the legislative campaigns of the Grange, the order was often caught in a paradoxical web of conflicting ideologies. On the one hand, grangers were true Jeffersonians clinging to the great Virginian's dictum that man's engagement in agriculture represented earthly communion with Nature; on the other hand, they were Hamiltonians asking for extensions of the arm of government into new spheres of activity. One instinct told Patrons of Husbandry to extol the virtues of farming and the glorious ideals of agrarian society; a countervailing impulse directed grangers to seek a greater share of governmental influence and participation in behalf of the citizenry. In essence, therefore, grangers viewed farming as the most noble calling of all; but beyond even this fact, they believed in active government—one that would direct and regulate and one that would not hesitate to solve problems as they arose.[22]

Granger inconsistency showed itself clearly during the grasshopper plagues of the 1870s. The preachments of Jefferson taught members not to expect relief from the government, but the ideas of Alexander Hamilton alerted them to the possibilities of enlisting aid from the government. In the showdown between the two philosophies, the needs of the day prevailed. Assistance in the form of surplus War Department blankets, shoes, and clothing was requested for distribution among locust victims.[23]

While governmental aid and sympathy were sought for some sufferers in society, little compassion was exhibited for criminals and vagrants. According to numerous chapters, individuals found not working should be prosecuted under provisions of vagrancy laws and then be made to pay society for their inactivity with service on public

[21] Indiana State Grange, *Proceedings*, 1880, p. 31; Illinois State Grange, *Proceedings*, 1895, pp. 41–42; National Grange, *Proceedings*, February, 1875, p. 91; 1877, pp. 121–22; 1895, p. 198; Virginia *House Journal* (1875), 164; (1876), 153, 175, 383, 420, 471–72; (1877), 226; Virginia *Senate Journal* (1876), 139, 230, 291; Maryland *Senate Journal* (1876), 99; Richmond *Virginia Patron*, February 16, 1877; Virginia State Grange, *Communication . . . Expressing Their Opinions and Wishes in Reference to the Inspection of Tobacco* (Richmond, 1876).
[22] Wisconsin State Grange, *Proceedings*, 1875, pp. 56, 80–81.
[23] *Ibid.*

works projects. On the question of what should be done with con-
victed murderers, grangers divided. Many members adhered to an
eye-for-an-eye concept of justice, while a small faction stressed reha-
billitation and forgiveness. Champaign Grange articulated the latter
position as well as any grange. The Illinois chapter called capital
punishment "a relic of barbarism" and asked for its abolishment.[24]

For the period under study, many of the legislative suggestions
made by grangers challenged old accepted ideas and customs and thus
proved too controversial to gain acceptance. The best case in point
involved the order's role in the female rights' movement. From the
beginning when Oliver Kelley's niece induced him to give women an
equal place in the organization, the order tested the leadership talents
of ladies and concluded from their performances that society did not
necessarily operate best with men doing the talking and with women
doing the listening. Results of the Grange's experimenting must have
been satisfying because the order never wavered on the question of
extending suffrage and equal rights to the fairer sex. To obtain that
goal, the farmers' brotherhood frequently joined established female
rights' groups in demanding that legislatures give women the same
status before the law as men and female teachers the same remunera-
tions as their male counterparts.[25]

With few exceptions, grangers were dedicated to the concept of
democratic or popular government. To them, the full citizenry should
be entrusted with opportunities to choose their representatives. Particularly objectionable were procedures in use for picking senators,
presidents, and vice-presidents. To grangers, making state legislatures
and electoral colleges responsible for these tasks was not good democratic practice; a better approach would rest upon direct elections and
pluralities. In essence, grangers were saying that the will and mandate
of the populace should be made supreme in determining who should

24 Maine State Grange, *Proceedings*, 1897, p. 62; Maryland *Senate Journal* (1876),
112; Indiana State Grange, *Proceedings*, 1880, pp. 26–27; 1897, pp. 39–40; Richmond
Virginia Patron, February 16, 1877; Indiana State Grange, *Proceedings*, 1880,
pp. 26–27; Alabama State Grange, *Proceedings*, 1874, p. 25; Champaign Grange,
Secretary's Minutes, February 7, 1876.
25 Colorado *House Journal* (1879), 118; California State Grange, *Proceedings*, 1890,
p. 98; Wisconsin State Grange, *Proceedings*, 1875, p. 84; New Hampshire State
Grange, *Proceedings*, 1874, p. 34; Ohio State Grange, *Proceedings*, 1891, pp. 28–29,
35; 1897, pp. 46–48; Vermont State Grange, *Proceedings*, 1891, p. 47; New Jersey
State Grange, *Proceedings*, 1891, pp. 55–56; Indiana State Grange, *Proceedings*,
1881, pp. 29–30; 1889, p. 45.

represent the people. Then to guarantee free elections and to mini-
mize voter intimidation, grangers asked for employment of the Aus-
tralian or secret ballot for all elections.[26]

In an era when wholesale corruption went largely unnoticed, the
Grange pleaded for and defended honesty as the best governmental
policy. To guard against the hiring of floaters, the bribing of voters,
and the buying of legislators, the National Grange directed its subordinate chapters to be champions of fairness and legitimacy and to
"make strenuous effort to secure in each State, the enactment of an
adequate law for the prevention and punishment of bribery in the
caucus or in the elections, and in the halls of legislation." Related to
these points was an implication by grangers that the nation was not
obtaining the best possible service from its elected officeholders. The
problem was simple; incumbents spent too much time seeking reelection. To relieve the situation, members of the Grange suggested
that one-term limitations be placed on offices and that tenures be extended. Such changes would prevent "them (officials) from spending
and neglecting the interests of the people and their duties as public
officers." [27]

There were other things connected with the operation of government which disturbed grangers. Most certainly, the Salary Grab Act
of 1873 left a lasting impression on their political outlook. Brazen self-
ishness shown by Congress in passing such an act in the midst of a
major depression produced a wave of indignation. A feeling developed among the citizenry that a general reduction in salaries and fees
given to or charged by governmental officials was in order whenever
the country found itself in depression. For the Grange, the idea of
adjusting salaries of public servants to reflect the business cycle and
of asking them to vote pay cuts to cover their alleged mistakes were
as applicable during the famine years of the 1890s as during the hardship period of the 1870s. But as earler, politicians paid little heed to
order demands for salary reductions. Legislators simply were not disposed to prune their incomes nor were they willing to cut those of
others in public service.[28]

[26] Social Grange, Secretary's Minutes, December 6, 1891; Illinois State Grange,
Proceedings, 1889, p. 32; 1891, pp. 57–58; Wisconsin State Grange, Proceedings,
1896, p. 37; California State Grange, Proceedings, 1890, pp. 60–61; Ohio State
Grange, Proceedings, 1891, pp. 28–29; 1897, pp. 46–48.

[27] National Grange, Proceedings, 1886, p. 120; Illinois State Grange, Proceedings,
1894, p. 50.

[28] New Jersey General Assembly Minutes (1897), 138, 689; Missouri House Journal
(1874), 113; Virginia House Journal (1877), 168; California House Journal (1874),

A few less spontaneous legislative demands were also made by grangers. Notable requests were presented in Kentucky and Minnesota. In the latter state, members of the Winona County Council of the Patrons of Husbandry expressed dissatisfaction in 1874 with the unfair structuring of their county board of commissioners. The body supposedly was not providing rural residents with the kind of representation they desired, so grangers presented the state general assembly with a petition asking action on their plan for reconstituting the board. Patrons of Husbandry noted that improvement would follow if the board's membership were revised so it would consist of the chairman of the boards of supervisors of each town in Winona County. Grangers from Barren County, Kentucky, discovered that legislatures in the 1870s showed little sensitivity to the will of the people. They sent their lawmakers a remonstrance pointing out that the trials of individuals accused of misdemeanors would be handled more efficiently by county courts. This suggestion, like dozens of others offered by countless granges, was read without subsequent discussion and then laid quietly to rest.[29]

In fairness to the elected officials of the nineteenth century, such brusque treatment was not always the rule. The 1875 session of the Virginia General Assembly was one that paid attention to suggestions offered by granges. For instance, when Farmville District Council of the Patrons of Husbandry asked the state legislature to pass a law forbidding allowance for counsel in certain cases, its recommendation was not read and forgotten. Instead, it was incorporated into House bill 256. The bill cleared the lower chamber and went to the Senate Committee for Courts of Justice, where it ran into a temporary snag. Committeemen failed to see any merit in the bill, so they returned it to the Senate without endorsement. But a majority of upper-chamber members used their own judgments, and took positive action on the proposal. This victory won by the Farmville group demonstrated the consideration expected of lawmakers by order members when they pushed for passage of specific legislation.[30]

379–80, 780, 851; Kentucky *House Journal* (1878), 932; Kansas State Grange, *Report of Historical Committee*, 6; Washington *House Journal* (1891), 145, 147; (1893), 482; Michigan *House Journal* (1874), 95; Kansas *House Journal* (1877), 659–60; Ohio State Grange, *Proceedings*, 1891, pp. 28–29; 1897, pp. 46–48.

[29] Washington *House Journal* (1893), 482; Kansas *House Journal* (1877), 659–60; Minnesota *Senate Journal* (1874), 238; Kentucky *House Journal* (1876), 560.

[30] Virginia *Senate Journal* (1875), 194, 401, 415–16, 432.

Yet to point an accusing finger at legislatures for not treating every suggestion received from granges with the same consideration shown the request made of the Virginia General Assembly in 1875 would be grossly unfair. In assessing blame for legislative inertia and for indifference of public servants themselves, consideration must be given public opinion in the late nineteenth century. Cautious and conservative were words which well described Americans. Most citizens were not favorably disposed to accept the views of grangers, feminists, single taxers, and laborites. Since the general mood of the nation trailed the lead taken by progressive pressure groups for legislative solutions to societal problems, lawmakers naturally felt compelled to proceed slowly.

Early experiences with legislators should have taught grangers something of the true mood of the nation and of lawmakers, but they did not. Grangers' insight into the workings of legislative bodies was never acute. Grangers were neither wise enough to concentrate their efforts in behalf of a few key programs nor patient enough to wait before demanding action on less important matters. As a result, granges continued to submit impressive-looking petitions whose demands ran the gamut from suggestions for laws covering liens and patents, arbitration procedures for lawsuits, and curbs on corporate charters to proposals calling for legal action against stock watering and public contributions to fairs. With so wide a spectrum of requests and with so hostile a public opinion, lawmakers really had little choice but to respond in a negative way. In nine of ten cases, a reading was all the attention given the order's lengthy petitions. Grange remonstrances often carried as many as twelve distinct proposals for legislation. Had one request been introduced at a time, legislators might well have looked with greater favor on the grangers' ideas.[31]

By firing repeated and scattered volleys at legislatures, grangers tended to obscure their positions on vital issues. The order's changing stand on immigration offers a good case in point. Lost in the shuffle of petitions was the National Grange's move toward nativism. At first only western grangers expressed much concern about the rising tide of immigrants, and in the main their interest was limited to the Chi-

31 Michigan *House Journal* (1874), 95; Washington *House Journal* (1891), 37, 145, 147, 174–75; Kansas *House Journal* (1877), 659–60; Georgia State Grange, *Proceedings*, 1873, pp. 8–9, 11–12; Columbus (Miss.) *Patron of Husbandry*, August 9, 1877; New York State Grange, *Proceedings*, 1895, p. 126; Illinois State Grange, *Proceedings*, 1891, pp. 57–58; West Virginia State Grange, *Proceedings*, 1895, pp. 21–22; Richmond *Virginia Patron*, February 16, 1877.

nese pouring into California. Had Congress closed the country to Oriental immigrants, California grangers along with many other racially minded residents of the Bear State would have had their desires on the subject satisfied.[32] The National Grange demonstrated little enthusiasm for nativistic appeals until the order came to a watershed period in its history. In 1888 when it still was not clear whether the Grange was on a threshold of rebirth or a brink of collapse, the parent body recommended a comprehensive four-point program aimed at preserving "America for the Americans." It was based upon the following assumptions:

> Each year our shores are being more and more crowded with undesirable citizens, which include adventurers, criminals, serfs, and the contaminated, lame and diseased of every character. Already our country is reaping the rewards of its folly, when it threw wide its gates, and became the "asylum for the oppressed of every land." Instead of being the desirable port it was intended to be, it has become the "dumping-place" of the Goth, Serf, and Vandal of every clime, and daily a harvest of anarchy is being reaped with all its attendant evils. With the present rate of increase of foreign citizenship, only a few decades will pass before America will be de Americanized, and the form of government and the institutions of our fathers will be over thrown.[33]

Prospects for a mongrelized United States worried grangers so much that the order sought to alert Congress to the possibility that the nation would become a hodgepodge. According to the farmers, something should be done immediately to see that adequate precautionary measures were taken to "protect the country from the introduction of undesirable citizens" who lacked "visible means" of support, good morals, healthy bodies, and peaceful intentions. Since rigid procedures for screening naturalization candidates had not yet been established, the National Grange suggested a possible course of action for dealing with the problem. According to the national body's plan, aspirants for citizenship would be judged fit or unfit solely on two grounds set by law. To pass, citizenship candidates would be required to establish a three-year residency record and to score satisfactorily on a standard literacy test.[34]

Tariff reform was another demand voiced by grangers. By its pas-

[32] National Grange, *Proceedings*, 1877, p. 132; California State Grange, *Proceedings*, 1886, p. 17.

[33] National Grange, *Proceedings*, 1888, pp. 125–26.

[34] *Ibid.*; *Congressional Record*, 50th Cong., 1st Sess., 183–84, 284. The Wisconsin State Grange took up the same theme in 1891. See Wisconsin State Grange, *Proceedings*, 1891, p. 6.

sage of the Morrill Tariff Act in 1861, Congress had moved firmly into the protectionist camp. More than two decades passed before the Grange had sensed sufficient danger to cause it to voice strong opposition. One of the first protests came in 1882, when the National Grange tried to block efforts of American fertilizer producers to induce Congress to impose heavy tariffs on imported additives. Members of the fraternal order's highest council charged correctly that the manufacturers' proposition would affect farmers adversely since it would place a direct tax upon them by advancing fertilizer prices. Apart from the 1882 appeal of the National Grange, there were other instances in the 1880s in which granges seriously concerned themselves with tariff issues. One involved the wool growers in the Wisconsin State Grange, who reacted angrily at the 1883 session after Congress had removed the ad valorem rate of 10–11 percent provided by the Tariff Law of 1867 on imported wool. The Wisconsinites disapproved and forwarded to Washington two resolutions asking for a return to previous levels.[35]

Tariffs became the focus of attention again in 1884 when Lecturer William H. Earle of the Massachusetts State Grange made the issue the major point of his annual address. He accused farmers of being grossly ignorant of the whole matter and pointed out that the level of American agricultural exports was dependent upon several factors, including the cost of consumer goods used by farmers. If protection were offered certain industries, prices of home-made products would increase accordingly, thereby adding to general inflation. In given situations where these rules were applied, observed Earle, farmers lost because a reliance upon exports was of greater benefit to them as a class than gains made possible by an artificially created home market buoyed by high duties. Recognition of these facts demonstrated a degree of sophistication, but Earle's knowledge of the situation did not excite many members in the order. Apathy continued, with no movement developing to shift the Grange to the side of free trade.[36]

Similarly, the efforts of Secretary of State James G. Blaine to forge a spirit of Pan-Americanism and to unite the United States and Latin America in a customs union aroused little interest among grangers.

[35] National Grange, *Proceedings*, 1882, pp. 83–84; F. H. Taussig, *The Tariff History of the United States* (New York, 1931), 239–41; Wisconsin State Grange, *Proceedings*, 1883, p. 48.

[36] Massachusetts State Grange, *Proceedings*, 1884, pp. 34–35.

Of the several state bodies existing in 1889, only Illinois expressed much concern about the probable effects of hemispheric cooperation on farmers. The thought of having reciprocity replace duties greatly frightened the Illinois chapter's committee of agriculture. Since both Americas were principally agricultural continents, the concern was understandable. Members of the committee believed that imports of Latin foodstuffs would depress the price of American farm products "without affording any corresponding benefit in our purchase of manufactured goods." Unfortunately for better hemispheric understanding and development, most individuals with positions of authority possessed no greater foresight; Blaine's call for blanket congressional authorizations to make necessary arrangements with all concerned fell on deaf ears.[37]

Grange interest in tariffs picked up considerably following release of a National Grange report written in 1890. It served notice on Congress that in tariff policy the order wanted farmers to "receive more and fairer consideration than has heretofore been accorded." To show that grangers understood tariff issues, their statement of 1890 mentioned the case of hides versus finished leather goods. Specifically, the parent body asked why foreign raw leather was allowed to enter the country duty free while finished products processed in the United States were granted tariff protection. Since no acceptable explanation was possible, grangers investigating the matter deduced that "manufacturing industries should [not] be afforded ample protection, when the producer of the raw material is forced to abandon an important industry because he is not protected." [38] The idea of Congress bestowing favors to special interest groups never set well with grangers, and the one-sided partiality shown in determination of tariff schedules proved to be no exception. In a pointed warning, the National Grange notified Congress in 1890 that it no longer planned to remain silent while the interests of farmers were abused by unfair treatment such as that accorded domestic producers of rawhides: "The time to turn down with impunity, the agricultural interest of this country has gone by. Henceforth we shall watch as well as pray. The quiet submission to neglect and unfair discrimination which has charac-

37 Illinois State Grange, *Proceedings*, 1889, pp. 60–61; *Alice Felt Tyler, The Foreign Policy of James G. Blaine* (Minneapolis, 1927), 165–90.
38 National Grange, *Proceedings*, 1890, pp. 90–92. For additional proof, see Illinois State Grange, *Proceedings*, 1891, p. 58.

terized the farmers in the past, has given place to a quiet but firm determination to know our rights and . . . to maintain them by every legitimate means within our reach." [39]

The New York State Grange carried the demands of the parent body one step further. The state chapter's committee on agriculture specified to Congress what rates it wanted for farm products:

PRODUCT	DUTY per UNIT		
barley	$0.20 per bushel		
buckwheat	0.20	"	"
beans	0.50	"	"
eggs	0.05	"	dozen
hay	5.00	"	ton
barley malt	0.40	"	bushel
	(of 34		
	pounds)		
peas	0.50 per bushel		
potatoes	0.30	"	"
dress poultry	0.20	"	head
hops	0.65	"	pound[40]

Perhaps the most imaginative tariff proposal to come before the Grange for consideration was a scheme devised by David Lubin. It was introduced at the 1894 session of the National Grange by the California and Illinois state granges. The Lubin Plan was unique for two reasons. First, it made the federal government responsible for the plight of farmers. Since few individuals recognized the full potential of the federal government for solving economic problems, Lubin was ahead of his time in that he possessed a partial understanding of possibilities. Second, the plan recognized that tariffs imposed for the benefit of industrialists were unfair and detrimental to farmers and to the general economy and welfare. Lubin's reasoning led him to surmise that revenues received from tariffs might be funded and offered to exporters as bounties. To some degree, the Agricultural Adjustment Act of 1933 provided a program similar to that envisaged four decades earlier by Lubin in the sense that the "dumping principle" was used as a means for increasing prices.[41]

---

[39] *Ibid.*   [40] New York State Grange, *Proceedings*, 1890, p. 108.
[41] National Grange, *Proceedings*, 1894, pp. 168–80; Illinois State Grange, *Proceedings*, 1894, pp. 23, 27–28.

Crucial to an understanding of the Lubin Plan is a grasp of the conditions at the time when Lubin was interpreting the national economic scene. Agricultural exports had provided foreign exchange since the early days of the Republic. Therefore, to conclude as Lubin had that great value should be attached to agricultural products and that special consideration should be given them was not illogical. Imports had to be paid for and interest on foreign loans had to be met either in bullion or in commodities. Since reliance on bullion was not practical, countries made most of their payments in goods. For nations in the same position as the United States, observed Lubin, the latter solution also had limitations. Protective tariffs added to the cost of foreign-made manufactures, increased prices generally, upset normal trade patterns, and rendered impractical the exportation of finished goods produced in developing nations. Because of these factors, Lubin noted, agricultural staples were left to constitute the great bulk of American exports.[42]

Although such an outflow of agricultural goods relieved the domestic market of surpluses, Lubin still observed serious drawbacks in the system for the American farmer. The maximum price obtainable for American agricultural products was no higher than the lowest price at which they could be purchased in the world, so in effect these goods were selling for export at the world's free-trade rate. With export and domestic prices for staple agricultural products being essentially the same, Lubin reasoned that it followed that these goods were selling at home and abroad at the world's free-trade or Liverpool price. Moreover, to make matters worse for agricultural exporters, transportation costs from the place of origin to England were automatically subtracted, whether in fact the goods were exported or sold in domestic markets. In summary then, this granger concluded that because of the protective tariff, manufactures were selling in the United States at artificially high prices, whereas agricultural products were going for export and home consumption at the world's free-trade price, less transportation costs from points of production to England.[43]

Lubin's proposals were predicated on another advanced assumption, namely that farmers deserved subsidization because they had been charged with the task of holding the nation's economy above water. Since the importance of the staple agricultural industry exceeded that of manufactures in value for world-trade purposes and was the only great industry whose goods sold at the world's free-trade

[42] National Grange, *Proceedings*, 1894, pp. 168–80.   [43] *Ibid.*

prices, farmers alone paid the cost of protection to manufactures.[44]

From this point, Lubin led directly into the next link in his chain of conclusions underlying his demand for new governmental policies. "Protection to manufactures," he observed, "made a high wage possible, this wage brought skill, skill developed inventive genius, and inventive genius produced labor-saving agricultural machinery." Having granted protective tariffs this much, the granger proceeded. "These machines in the hands of the American producer of agricultural staples amply repaid him for any cost for the protection of manufactures. With the powerful aid of labor-saving machinery, he could, until recently, produce his crop so profitably as to enable him to compete successfully with the cheapest labor in the world." But this situation no longer prevailed.[45] Lubin advised individuals who sought agrarian support for high tariffs by pointing to benefits once accruing from such duties to recognize that the edge had vanished. Lubin theorized that the American producer of 1894 no longer possessed past advantages because of a new global trend of increasing profits through use of the world's cheapest field labor and sophisticated machinery. A result of ensuing losses of supremacy would materially reduce American exports of staple agricultural products and eventually would remove the prop which had been a stabilizer of the whole protective system. A collapse of American agriculture would leave only two alternatives open to the United States as a nation. Protection for manufactures would have to be abandoned, or the source of such aid would have to be shielded.[46]

Lubin firmly believed also that there had to be a change at one end or at the other end of the trade nexus, and he confidently expressed the hope that he had found the best solution. This westerner hypothesized that either prices of industrial wares would have to be decreased to the world's lowest free-trade levels or that prices of agricultural staples would have to be enhanced by some means other than a protective duty. A tariff alone would not help much, Lubin predicted, since a duty could not improve the domestic price of an export which was being sold at the world's lowest price. A solution to the dilemma lay only in having the government give bounties to domestic growers who exported foodstuffs and fibers. Lubin's presentation concluded with a suggestion for funding the plan equitably. Care should be taken, Lubin ventured, in meeting the cost of providing a bounty system; the federal government could not be placed in a position of taxing

44 *Ibid.*   45 *Ibid.*   46 *Ibid.*

one group to benefit another. But if incomes collected from imposts were used as the principal source for the bounty fund, no one could justifiably accuse the government of partisanship or wrongdoing.[47]

When presented to the National Grange for consideration, the Lubin Plan was already well known in grange circles and was well defined. Nevertheless, as a matter of routine, the general body still sent the scheme to the Committee on Agriculture for further study and assessment. From this body came three divergent reports. The majority report rejected the Lubin Plan, dismissing it as "clearly unconstitutional." The majority recommended only that the National Grange continue to pursue its policy of seeking high protective duties on wools, farm products, hides, and pelts. Master John B. Long of the Texas State Grange filed one of two minority reports. Higher tariffs were the antithesis of what the neo-Calhoun Democrat from the Southwest wanted. To this maverick, what was needed were not duty increases and bounty schemes, but rather across-the-board adjustments to insure that all future international commercial negotiations by the United States would be made with both agricultural and industrial interests in view. Another dissenting brief came from A. P. Roache of California. His report was a short endorsement of the Lubin Plan and represented the views of the California and Illinois state granges. It contained nothing new, but it did summarize again why the National Grange should act favorably on Lubin's resolutions.[48]

Much bitterness erupted as an aftermath of the action taken by the general assembly of the National Grange on the three reports. Roache's minority report was rejected, thirty-two nays to seventeen yeas. A regional breakdown showed the West solidly in favor of the Lubin Plan, the East steadfastly against it, and the middle sections evenly divided. Long's report was defeated by an even greater margin. Figuring after the first vote that compromise would be better than nothing, the Texan submitted five amendments to his original report and asked for a second tally. When these efforts produced no voting shifts, he asked to be relieved from "further service on committees during the session." His request was granted, and the obviously shaken Texan never attended another session of the National Grange. In fact, the Lone Star State did not seek representation at the next meeting.[49]

Although the National Grange had failed to sanction the Lubin

47 *Ibid.*   48 *Ibid.*, 168–76.   49 *Ibid.*, 176–80.

Plan in 1894, efforts in its behalf were not abandoned. David Lubin spent most of 1895 on the road, traveling and explaining his scheme from California to Massachusetts. In the end, however, his determination paid few dividends. Easterners continued steadfast in their opposition. For them, reciprocity treaties and protective tariffs could accomplish more than any bounty scheme. As a result of the East's inflexibility, the National Grange never approved Lubin's resolutions and never deviated from its reliance on traditional courses of action for solving international trade problems.[50]

Something similar to the Lubin Plan had been proposed by Bristol Grange No. 53 at an 1879 session of the Wisconsin State Grange, but this proposal received little publicity. By the Wisconsin Plan, the president would be authorized by Congress to send American grain to starving peoples of the world. Money for this assistance would come from special legal tender notes issued by Congress. In exchange for aid, nations receiving American food would be expected to pay interest to the United States at rates comparable to those earned on United States bonds.[51]

The only other time in the century when granges deviated from their usual stand on the question of tariffs was after the Spanish-American War. Believing that the United States might be ruined by cheap agricultural imports from the islands acquired from Spain, the Massachusetts State Grange in 1898 and 1899 asked the federal government to take protective action. Specifically, these Bay Staters wanted the welfare of American producers insured; they did not want island goods being imported duty-free into the United States.[52]

On another important issue of the late nineteenth century, governmental monetary policy, grangers were less radical and dogmatic than several other agrarian groups. In the mid-1870s, leadership of the National Grange did not share the Greenback Party's view that the Treasury Department should issue more paper money. In fact, many grangers insisted that the government recall circulating greenbacks. Their speeches on fiscal subjects often reflected opinions which were closer to those expected of bankers than from debtor farmers. Read, for example, Master Dudley W. Adams' annual address delivered in 1875. His words sounded more like those found in the pages of *Poor*

---

[50] *Ibid.*, 1895, pp. 146–47, 162–63; New York State Grange, *Proceedings*, 1895, p. 170; West Virginia State Grange, *Proceedings*, 1895, pp. 21–22.

[51] Wisconsin State Grange, *Proceedings*, 1879, p. 64.

[52] Massachusetts State Grange, *Proceedings*, 1898, pp. 83–84; 1899, pp. 75–76.

Richard's Almanac than in a speech presented to a gathering of nineteenth-century farmers:

> A thousand years ago learned and thoughtful chemists devoted the energies of a lifetime to a vain search for the wonderful philosopher's stone whose magic touch should convert the baser metals into purest gold, and thus fill the whole world at once with wealth and luxury. Today we have numerous citizens who are eagerly pursuing the same phantom. They are torturing their poor brains to devise some plan whose talesmanic [*sic*] power will transmute bits of printed paper into countless millions of actual money of such a subtle nature that, true as the needle to the pole, it shall go straight to the pockets of the poor, and, like a veritable "will o' the wisp" forever evade the clutches of the rich. It is an indisputable fact that our country is now seriously suffering from a derangement of finances. We need not be at loss to know the cause. It is a solemn reality that our country has passed through a most wasting civil war. It cost us in money, time lost, industry disturbed, material destroyed, [and] production stopped. . . . To bridge over this emergency of the hour the government issued great volumes of irredeemable paper currency, which we used as money, and thus for a time disguised and hid our poverty. By using this currency our judgment of values became more and more confused as we drifted farther from the world's standard. We totally failed to realize our changed circumstances and to inaugurate a corresponding system of economy and industry, and consequently, with an inheritance of debt, extravagant habits and distorted judgment of values, we have been necessantly drifting to leeward. Out of this trouble there is no royal road. Only by a return to habits of industry and economy, guided by intelligence, can we regain our wealth and remove our load of debt. As an auxiliary to this, we want a stable and sound currency that shall be a reliable measure of values and recognized as such by all the civilized world.[53]

It was not until the late 1880s and the early 1890s that widespread opposition to the National Grange's conservatism in monetary affairs began to be heard. A prelude to what was to transpire nationally unfolded in 1888 when the Colorado State Grange announced that it now was favoring looser monetary policies. One year later, rebellious Turks in the Grange took advantage of favorable conditions and overturned Old Guard conservatives. Beginning in Illinois and terminating at the 1889 session of the National Grange, "radical" grangers seized initiative and pushed through resolutions favoring free and unlimited coinage of silver. The vote on the silver resolution was decidedly sectional in character; delegates from Connecticut, Maine,

---

[53] National Grange, *Proceedings*, February, 1875, 12–13. For a discussion of the Greenback Movement, see Irwin Unger, *The Greenback Era: A Social and Political History of American Finance, 1865–1879* (Princeton, New Jersey, 1968).

New Hampshire, New York, Rhode Island, and Vermont cast six of the nine negative votes. There were twenty-nine affirmatives. Into the next century, the National Grange and an increasing number of its subordinates supported free silver. Although there were some conspicuous holdouts, most chapters had been won to the cause by 1895. Even the conservative New York State Grange could be placed firmly in the bimetallic camp after its 1895 session.[54]

Before this conversion to silver among order members occurred, one of the most articulate conservative statements on national finances originating at a grange session came from a meeting of the Michigan State Grange. In 1891 members of that chapter released their evaluations of free-silver panaceas and of a proposition offered by Congressman M. D. Harst "to turn the people over to the tender care of banks." The latter plan to give banks exclusive power of issuing money was rejected on several grounds. First, Michigan grangers were "unalterably opposed to the issuing of money by either State or national banks, no matter how well secured or safely guarded" because they regarded "such issues as a dangerous surrender of the functions of government to private corporations." Second, these grangers rejected the inducement used by Harst to gain acceptance for his proposal. He had proposed that issuing banks be taxed at a rate of 2 percent per annum on their circulation. To the grangers assessing Harst, his whole plan was a fraud because it was obvious that such taxes would be absorbed by bank patrons and not by their stockholders.[55]

This expertise shown by the Michigan State Grange in analyzing demands of bimetallists for free and unlimited coinage of silver demonstrated the degree of understanding which some grangers had of the workings of international and national finance. Essentially, the opposition offered by Michigan grangers to the silver bugs' pleas was based on a belief that governmental reserves of gold and silver bullion were already more than sufficient to produce inflation, but that prevailing governmental policy prevented realization of the full expansion potential. These Michiganites pointed out that during Grant's presidency a choice had been made by the Treasury Department and

[54] Colorado State Grange, *Proceedings*, 1888, p. 19; Illinois State Grange, *Proceedings*, 1889, p. 41; 1890, pp. 26–27; National Grange, *Proceedings*, 1889, pp. 110–11; Ohio State Grange, *Proceedings*, 1890, p. 36; 1891, pp. 28–29; 1897, pp. 46–48; West Virginia State Grange, *Proceedings*, 1895, p. 22; Greene County Pomona Grange, Secretary's Minutes, May 29, 1891; Massachusetts State Grange, *Proceedings*, 1890, p. 79; New York State Grange, *Proceedings*, 1895, p. 126.
[55] Michigan State Grange, *Proceedings*, 1891, p. 37.

by Congress not to turn the nation to an inflationary course. That decision, not a lack of silver, explained why the amount of gold and silver notes in circulation never substantially exceeded the value of gold and silver specie in the treasury. It was no secret that legislative and executive leaders of both parties deemed it neither necessary nor beneficiary to have periods of inflation, and the Michigan State Grange recognized this fact.[56]

Unlike most chapters, the Michigan group understood that rising prices could be produced without resorting to free silver if inflation ever became a national goal. If only $100,000,000 in gold were capable of maintaining $346,000,000 in greenbacks, asked the Michiganites, why could not the same redemption-fund principle be applied to expand the volume of gold and silver certificates and notes in circulation? Instead of having treasury silver worth $395,000,000, gold value, supporting a like amount of silver certificates and treasury notes and instead of having treasury gold valued at $140,000,000 backing an equal sum of gold certificates, the nation might have existing gold and silver reserves holding up $484,000,000 in gold certificates and $1,366,700,000 in silver-backed issues. Moreover, by such a line of reasoning, Michigan grangers were brought to conclude that there was no special virtue in free-silver panaceas. In fact, rather than favorable results from an acceptance of free-silver proposals, the chapter's members envisaged vexing international problems, gold drains, and greater internal complexities accruing from applications.[57]

When compared with the soundings made in other areas, the voice of the National Grange was surprisingly silent on monetary questions. The Populists' panacea, free silver, never excited much interest among grangers. The order's range of activities was much broader, and its members' grasp of the complexities of the day was generally superior to that of Alliancemen and Populists. Twenty years experience as spokesmen for farm families had taught grangers to be ever mindful of the fact that there were few simple answers to the complex problems facing farmers.

For example, grangers consistently demonstrated more interest in practical tax reforms than in the wild inflationary schemes so popular with other agrarian groups during the late nineteenth century. On the state level, the demands most often made by granges were equalizations of land values for tax purposes and for tax deductions on mortgaged property. In regard to the latter request, grangers believed

56 *Ibid.*, 37–39.   57 *Ibid.*

it unfair to tax mortgaged land at the same rate as property held free of debt. To appreciate what grangers had in mind, it is necessary to know something of the debt problem confronting farmers in the late nineteenth century. A high percentage of farms had mortgages, and farmers struggled against declining prices to pay them off. Therefore, if the conditions are not placed in the shadows by the plan's impractical facets, it is easy to look beyond the grangers' naïveté to develop sympathy for their request.[58]

Throughout the late nineteenth century, grangers also gave priority to the eradication of tax inequities. This movement originated in 1874 when the Wisconsin State Grange distributed copies of an inflammatory circular by A. Gaylord Spalding which sought to arouse farmers to exercise their strength at the polls. The circulation of this leaflet was an indication of how far the state grange was willing to go in its fight for ending injustice in the collection of taxes. Almost every paragraph made reference to the fact that farmers were being bled of precious dollars by the enforcement of laws which had been enacted in the interests of society's "deadheads" by their hand-picked spokesmen in Congress and in state legislatures. The Ohio State Grange used a less dramatic approach in 1876 to register its complaints with the legislature. The Buckeye chapter simply called on state solons to equalize real-estate values for taxation purposes.[59]

Many of the demands first made in the 1870s were heard again in the 1890s. Patrons still believed that they were being victimized by unfair tax laws. In particular, their complaints focused on the way assessments were made on property. Granges alleged that many statutes favored the nonagricultural real-estate owner over the farmer. To end this form of discrimination, chapters continued to petition lawmakers for revisions on all standing laws pertaining to the topic. For the most part, these requests were simple cut-and-dried statements of general grievances. Occasionally, however, an extra dimension was added to make the farmer's case for tax reform even stronger. Illinois grangers bolstered their efforts in 1895 with statistical information showing that farmers were paying 79 percent of the taxes col-

[58] Iowa State Grange, *Proceedings*, 1873, p. 41; Ohio State Grange, *Proceedings*, 1876, p. 51; Vermont State Grange, *Proceedings*, 1875, p. 28; West Virginia *House of Delegates Journal* (1895), n.p. (January 25); Shannon, *The Farmers' Last Frontier*, pp. 184–90, 303–309.

[59] A. Gaylord Spalding, "Voices of the Grangers," an undated final draft of a circular which was to be printed, in Osborn Papers; Ohio State Grange, *Proceedings*, 1876, p. 88; New York State Grange, *Proceedings*, 1891, p. 120.

lected in Illinois from meager returns from a generally unprofitable profession.[60]

Unequal taxation was also the subject of a lengthy paper issued in 1892 by a special committee of the Massachusetts State Grange. The treatment afforded tax inequity was very sophisticated, considering that a group of nineteenth-century farmers were the authors of the report. In summary, the committee suspected that tax problems resulted more from a failure to adjust laws to meet new conditions than from an obvious conspiracy. Except for the manner in which taxes were assessed, the Massachusetts of 1892 bore little resemblance to the state in earlier decades. Real progress was evident in all but one area. Individuals with the most land still paid the most taxes regardless of other assets or liabilities. The basic laws governing the collection of taxes had not been changed substantially since the days when the size of a man's property holdings was a fair gauge of his wealth. Considering that the old criteria for measuring affluence had long ceased to be reliable, a reassessment certainly appeared appropriate.[61]

The whole subject of tax collecting needed to be studied to ascertain what new sources of revenue could be tapped. From its research, the state grange felt confident that much would be gained if the general assembly looked into the merits of raising money through a doomage law. The idea for such state legislation was not new; the results were good enough to lead Massachusetts grangers to conclude that their state was in need of such an enactment. Coupled with this demand was the connection that additional funds could also be obtained through revocation of special privileges accorded holders of municipal bonds. For eleven years, interest earned on these investments had been legally tax exempt. Although nothing could be done about special rights already given, the Massachusetts State Grange believed injustice would continue to prevail if steps were not taken immediately to insure procurement of taxes from future issues.[62]

For these conscious tax reformers in the Grange, equity in taxation had a broad meaning. By their definition, every holder of property and every recipient of income should be subject to the same tax rates. There should be no exceptions. Being a purist, the granger even wanted Indian lands, railroad properties, and real estate owned by

---

60 Illinois State Grange, *Proceedings*, 1895, pp. 51–53. For additional insight, see 1891, pp. 57–58; Ohio State Grange, *Proceedings*, 1891, pp. 28–29; West Virginia State Grange, *Proceedings*, 1895, p. 22.

61 Massachusetts State Grange, *Proceedings*, 1892, pp. 131–44.

62 *Ibid.*

churches assessed and taxed. But above all, he wanted antiquated tax laws removed from the statute books, so farmers' property no longer served as the major source of tax revenue.[63]

Some indication of how important tax reform was to grangers may be seen by examining the proceedings of the Pennsylvania House of Representatives. The Pennsylvania *House Journal* for 1895 mentions no fewer than forty-three petitions sent by granges to inform lawmakers of their support for a revenue bill before the general assembly. Almost as much enthusiasm developed in 1899 after the introduction of a bill entitled "An act to provide revenue by taxation and to secure greater uniformity of taxation." Proffered by Representative William T. Creasy, this bill gained a favorable endorsement from the House Committee on Law and Order and passed the lower house before encountering opposition in the Senate Committee on Finance. There, strong objections raised by the conservatives who controlled the body were enough to prevent the bill from reaching the Senate floor in time for a vote. Although special interests had won another victory at the Grange's expense, the backing given the bill by Pennsylvania grangers was still noteworthy. In the house, none of the negative votes recorded against the bill came from representatives of counties containing granges which had petitioned the legislature in behalf of passage.[64]

On the national scene, grangers were involved with two tax proposals. One was the income tax. By a vote of twenty-four to seventeen, the 1880 session of the National Grange approved a resolution calling for the "enactment of a graduated income tax to the end that all wealth may bear its just and equal proportions of the expenses of government and that productive industry be so far relieved from the burdens of taxation as shall be consistent with strict justice to all." Subsequent resolutions were adopted and forwarded to Washington until the Fifty-Fourth Congress enacted a tariff law with an income-tax rider. Passage greatly pleased grange leaders. They began praising the new revenue measure, and they warned members to be prepared to defend it against all possible criticisms which might be leveled by conservatives and reactionaries. The 1895 decision of the Supreme Court de-

---

[63] Social Grange, Secretary's Minutes, December 6, 1891; Ohio State Grange, *Proceedings*, 1876, p. 88; Kansas *Senate Journal* (1874), 269; Illinois State Grange, *Proceedings*, 1891, pp. 57–58.

[64] Pennsylvania *House Journal* (1895), 1370–73, 1719, 1807; (1899), 794, 1063, 1363–65, 1374–75, 1411, 1464, 1559, 1620, 1674, 1722, 1802, 1829, 1951; Pennsylvania *Senate Journal* (1899), 1799.

claring the income-tax law unconstitutional was far from popular among grangers. Reaction in Illinois was especially bitter. There, the state grange resolved that farmers should not pay any taxes because all gains represented income. Had not the highest court in the land ruled that income taxes were unconstitutional? Although the resolution sounded militant, the call for a tax boycott failed.[65]

Henry George's controversial scheme for a single tax was another proposal of national interest. Briefly, George and his disciples in the Single Tax League favored general abolition of all taxes except for those levied on land values, a proposal that generated little enthusiasm among grangers. In fact, the National Grange and many of its subordinates stated in no uncertain terms that they disapproved of the entire single-tax movement. Master Milton Trusler of the Indiana State Grange expressed the feelings of many members when he labeled the single taxer a "would-be robber of our homes." The Hoosier added that adoption of George's nostrums would relieve no one of poverty.[66]

To secure passage of desired legislation, grangers relied on an assortment of methods. In most cases, they simply presented their demands in the form of resolutions, petitions, and memorials. On occasion, however, chapters employed lobbyists to "educate" lawmakers. One grange was even successful at forming a minor farm bloc. In 1888 Massachusetts state grangers joined representatives of independent farmers' clubs and creamery associations to form a united front for the representation of agrarian interests at the Massachusetts General Assembly and for the fight against oleomargarine. Even though the state grange had been successful in uniting antioleomargarine factions and in presenting petitions to the legislature, state solons retained enough of their independence to defeat the oleo bill proffered by the alliance. Fortunately for the bloc, however, all was not lost; part of its program was salvaged. The legislature amended a state dog law and appropriated more money for the state agricultural college at Amherst. No action was taken on a dairy commission bill. Thus, through cooperation with other groups and by galvanization of their own efforts,

65 National Grange, *Proceedings*, 1880, pp. 91, 93; Harold U. Faulkner, *Politics, Reform and Expansion, 1890–1900* (New York, 1963), 185; Ohio State Grange, *Proceedings*, 1894, p. 10; Massachusetts State Grange, *Proceedings*, 1892, pp. 143–44; West Virginia State Grange, *Proceedings*, 1895, pp. 21–22; Illinois State Grange, *Proceedings*, 1895, pp. 30–31, 53.

66 National Grange, *Proceedings*, 1889, pp. 108–109; Massachusetts State Grange, *Proceedings*, 1899, pp. 54–58; Indiana State Grange, *Proceedings*, 1887, pp. 8–10.

Massachusetts grangers gained acceptance for 50 percent of their program.[67]

Before any conclusions may be made concerning legislative aspirations of grangers, one point must be made. Essentially, it involved the order's attitude toward the United States and her system of government. Generally speaking, members of the agrarian brotherhood were proud and patriotic Americans; they certainly were not revolutionaries or syndicalists in the true sense of these words. Their pride in their nation had been revealed on too many occasions and in too many ways for them to be so classified. A granger's relationship to the United States differed sharply from that of a radical. For instance, Patrons of Husbandry displayed no revolutionary characteristics midway in the 1870s when funds for the Washington Monument were running low. Instead, they contributed generously in order that the project might be satisfactorily completed in time for the centennial of Independence Day. Few persons bent on destroying the nation would have paused to honor its heroes and its past. Patriotism of a different sort was demonstrated in 1898 after the United States found herself at war with Spain. From an average granger's perspective, there was no choice but to defend American involvement. In typical granger style, members never questioned whether the struggle was more than a fight between the forces of good and evil. As a result of this blind dedication and devotion, chapter after chapter pledged unqualified support to the American cause. No radical organization would have blindly backed a nation engaged in such a conflict as the Spanish-American War.[68] In retrospect, the average granger probably felt that there was no crime in washing the flag when it was dirty, but he no doubt deemed it treasonous to advocate burning Old Glory. To him, constructive criticisms and suggestions aimed at making the United States better were acceptable, but blatant attacks against the system and calls to arms against the nation were not.

Essentially then, grangers were conservatives in the sense that they did not call for any radical departures from the *status quo*. Moreover, the underlying presence of Jeffersonian agrarianism in the order's legislative proposals suggests that these activities were more reactionary than conservative. For most grangers, Thomas Jefferson's dream

[67] Indiana State Grange, *Proceedings*, 1893, p. 26; Massachusetts State Grange, *Proceedings*, 1889, pp. 72–77; 1891, pp. 23–25.

[68] National Grange, *Proceedings*, February, 1875, pp. 101–102; 1895, pp. 199–200; 1896, pp. 19–20; Illinois State Grange, *Proceedings*, 1895, pp. 31–32, 54.

of establishing a democracy with inviolable rights for farmers was still obtainable in the United States. Making grangers look even more reactionary was the fact that the past held more than just pleasant memories for them. It also served as their pattern for the future. Therefore, on the bases of Jeffersonian undertones and lapses into sentimental nostalgia, grangers have to be deemed reactionary. But to fail to explore the positive facets of grange legislative demands would deny justice to the grangers' efforts. A good case can be assembled to show that the farmers' order was basically a progressive organization. Members of the brotherhood usually faced the future without any real fear, and they generally believed in a larger role for government in the regulation of society. As a matter of fact, bigness seldom frightened them as long as the citizen was guaranteed certain safeguards and a role in governmental proceedings. In summary, it is safe to say that no label accurately covers the Grange's legislative concerns, but the word paradox describes the inconsistencies attached to the order's legislative activities.

How much the Grange gained by making its legislative ideas known is not ascertainable, but one fact is clear. The order's efforts suffered from diffusion. Like the Mad Hatter in *Alice in Wonderland,* grangers never seemed to be accomplishing much because they were always trying to move in all directions at once. Had the order concentrated its efforts and had it not pushed for everything at the same time, it more than likely would have asserted more pressure on lawmakers to act positively on its most important propositions. Their chances certainly would have been better for arranging alliances with groups having similar interests. Nonetheless, no such mastery of lobbying techniques occurred in the late nineteenth century. As a result, few laws can be linked directly to the efforts of grangers. If an enactment bore a resemblance to what the order desired, its passage was not necessarily due to the effectiveness of the Grange as a lobbying agent. Instead, it was more often a case of grangers supporting legislation coveted by others. In final analysis, the Grange aided the cause of much legislation but was solely responsible for little.

CHAPTER X

Communications, Transportation,

and the Grange

In 1967 to commemorate the National Grange's centennial year, the order sponsored for law students a contest that demonstrated well the viability of the Granger Law myth. According to the rules, winners were those who submitted "the best papers on the impact and significance of the Granger Laws and what they have meant in modifying and improving our capitalistic system." [1] The fact that in 1967 grangers were still linking their organization to the so-called granger laws is understandable if the word *granger* and its double meaning are kept in mind. But for a study of the farmers' organization begun by Kelley, granger should refer to a member of the Patrons of Husbandry; it should not mean any farmer involved in the general rural unrest of the 1860s and 1870s and to the agrarian movement for railroad regulation.

Given that limitation, grangers loomed even less important as a determining factor in the passage of state railroad regulations than George H. Miller suggested in his debunking studies of granger laws. According to standard interpretations, these laws resulted from militant farmers' demands for railroad legislation. Used loosely, the word granger is employed as a blanket term to refer to any farmer who battled for his rights. For a farmer to be classified a granger by this definition, membership in the Grange was not a necessary prerequisite. The questions raised and answered by Miller revolved around whether militant farmers were responsible for the railroad regulatory acts passed in Iowa, Illinois, Wisconsin, and Minnesota; he was not seeking to determine whether agrarian dissidents were or were not members of the fraternal order. While standard interpretations give agrarian interests credit for the controls directed against railroads, Miller stripped the word granger from the granger laws. He concluded that farmers have been too closely associated with railroad

[1] "Centennial Year Projects. . . ," *National Grange Monthly*, LX (November/December, 1966), 20.

regulations and that not enough importance has been accorded other groups. In fact, farmers do not deserve the credit they have received.[2]

In the case of Iowa, for example, Miller introduced three new inter-related facts bearing upon developments there. Within a decade, the direction of trade changed from a north-south axis to a west-east course, due partly to the Civil War; the Chicago and Milwaukee Railroad moved across the Mississippi River into the Iowa hinterland; and much resentment and hostility arose in river towns against railroads. Lengthening of the C & M and rechanneling of trade brought economic suffering to such towns as Burlington, Clinton, Davenport, Dubuque, Fort Madison, Keokuk, McGregor, and Muscatine because the handling of goods for shipment to points south had been the life-blood of these municipalities. With rail-carried commerce bypassing them, river-town grain dealers, warehousemen, millers, and packers did the natural thing; they reacted to protect themselves. Rallying local civic and business leaders to their side, they initiated determined campaigns to obtain railroad regulation from the legislature. Had the coming of the Grange not generated so much excitement, Iowa's granger law might have been called the rivertown law. Point and rate discrimination became illegal only after powerful nonagricultural blocs had deemed it desirable to have restrictions.[3]

As a result of Miller's pioneer work, the role of the Grange becomes clearer. Iowans who claimed membership in the farmers' brotherhood looked at railroad problems and at the campaigns initiated by river towns for controls and concluded that these efforts were worthy of support. The relationship of grangers to railroad regulations, therefore, was one of assistance; it was never one of predominance or initiative. But given that secondary role, grange leaders spoke out strongly. At the 1873 session of the state grange, Master A. B. Smedley devoted a major part of his annual address to railroads. Smedley's report was one of the strongest and most sophisticated cases for railroad regulations ever presented in Iowa by a member of the Grange during the First Granger Movement. The master oversimplified some complicated matters and he displayed the usual antirailroad biases,

[2] George H. Miller, "Origins of the Iowa Granger Law," *Mississippi Valley Historical Review*, XL (March, 1954), 657–80; George H. Miller, "The Granger Laws: A Study of the Origins of State Railway Control in the Upper Mississippi Valley (Ph.D. dissertation, University of Michigan, 1951); George H. Miller, *Railroads and the Granger Laws* (Madison, 1971).

[3] Miller, "Origins of the Iowa Granger Law," 657–80.

but these faults were overshadowed by his otherwise brilliant analyses.[4]

First, Smedley argued effectively for a change of the formula used to tax railroad properties. Using as the value of railroads the total assets per mile of track, he showed that common carriers were taxed at rates dramatically lower than those for farmers. Owners of agricultural land paid anywhere from 20 to 50 mills per dollar value of their property, while railway companies were assessed only four mills per dollar on their developed land holdings. He went on to say that their tax "was levied on the gross earnings per mile." [5]

Secondly, Smedley demonstrated that it would be constitutional for states to place controls on railroads. His case was based on three considerations. First, railway lines in Iowa were semipublic because they had received governmental donations totaling $33,473,093. With 3,642 miles of track in the state, "an average apportionment of the amount donated to all roads in the State ... would amount to $9,190 for every mile of road built, or a fraction over one-half of the actual cost of all our railroads." In effect, therefore, he was maintaining that the state should be able to govern that which it owned. Second, the Supreme Court had decided in the case of the *Sheboygan and Fond du Lac Railroad* v. *Fond du Lac County, Wisconsin*, that railroads were "public highways, like bridges, town roads, etc., and as such were subject to the same rules." Third, in the acts granting public lands to the various roads, there were clauses which gave the state the right to regulate freight and passenger tariffs. For these reasons, therefore, "there is no question," he said, "but what the Legislature of the State has full power to regulate the relation of the railroad to the people. It only needs the disposition to do it, and that they fully comprehend it and the subject in all its bearings." [6]

Next, Smedley used statistics to explain that point discrimination against river towns was adversely affecting farmers. He alleged that railroad pricing policies were "shutting ... [Iowa food producers] out from all advantages which might at certain seasons accrue ... from that great water route." There were times when St. Louis and New Orleans markets were better than those of Chicago and Milwaukee, but railroad officials preferred that agricultural products go to Chicago so that the carriers might have longer hauls. Tariff rates for first-

4 Iowa State Grange, *Proceedings*, 1873, pp. 28–35.
5 *Ibid.*, 29.
6 *Ibid.*, 30–31.

class freight per one hundred pounds showed clearly that the leading roads running east and west through Iowa were discriminating against shippers who sought to transport cargoes to river towns.

Chicago, Burlington & Missouri
Mt. Pleasant to Burlington, 28 miles .... $ .25
"   "   " Chicago, 218 miles .... .85

Chicago, Rock Island & Pacific
Des Moines to Davenport, 168 miles .... .60
"   "   " Chicago, 351 miles .... .75

Northwestern Railroad
Cedar Rapids to Clinton, 81 miles .... .40
"   "   " Chicago, 319 miles .... .75

Illinois Central
Waterloo to Dubuque, 93 miles .... .60
"   " Chicago, 300 miles 1.00

Milwaukee & St. Paul
Cresco to McGregor, 62 miles .... .40
"   " Chicago, 300 miles .... 1.05[7]

Smedley concluded his remarks with a review of desirable solutions to the railroad dilemma. Relief would come, he said, only if the federal government and the Iowa General Assembly enacted omnibus bills to insure better transportation services. To do less than to approve comprehensive measures would insure failure, he pointed out, because "no one of the measures suggested [heretofore] will bring ... the relief needed." Smedley then went on to explain that he was urging the state grange to call on the state legislature to prohibit various forms of discrimination against persons and places and to provide for the creation of a powerful board of railroad commissioners with authority to set rates. At the same time, he stated that the order needed to ask Congress to stop foreign investors from draining the American economy, to widen and dredge the mouth of the Mississippi River, to construct a Niagara ship canal, and to finance the establishment of a government-owned, double-track railroad line from the Atlantic Coast to the Missouri River. In regard to the latter point, Smedley said he favored the building of a government-owned railroad because

7 *Ibid.*, 33–34.

its operation might prove useful in determining whether ordinary lines were operating efficiently and equitably.[8]

After receiving a mandate from his state grange, Smedley prepared a brief for the legislature. In it, he explained why grangers were outraged at railroad practices and discussed what they were seeking in terms of legislation. Being firm and precise at the same time, Smedley left no doubt where the order stood on the question of railroad controls. Although Smedley's ideas prevailed in the state grange through 1873, the body nevertheless limited its demands to requests for "a law prescribing maximum rates for passengers and freight on the railroads of Iowa." By the end of 1874, however, the scope of activity had been broadened. Apparently, "wiser heads" in the chapter had come to realize how imperative it was to have all of Smedley's legislative suggestions adopted into law at once. This change of direction was first evident February 17, 1874, when grangers testified before the railroad committees of the state legislature. In substance, the type of law now being suggested provided for a board of railroad commissioners with power to supervise all aspects of intrastate railroading. On the other hand, Iowa grangers did not want a "cast iron tariff bill." They did not feel that the legislature had studied railroad problems long enough to handcuff a commission with fixed rates. "But the members of the General Assembly, lamented Chairman [John] Scott [of the grange legislative committee on railroads], 'entertained the most crude notions as to the whole matter, and grappled the subject in that spirit of innocence with which an infant would play with a serpent.' The result was the 'Granger Law'—of which leading Iowa Grangers disapproved." [9]

When compared with those in Wisconsin, Minnesota, and Illinois, the bill enacted into law in Iowa without the full blessing of the state grange stands as perhaps the most comprehensive railroad law approved by a legislature during the early 1870s. The Iowa railroad law of 1874 classified lines "according to the gross amount of their respective annual earnings within the state per mile" of track, established detailed freight-rate schedules, made discrimination of persons illegal, required roads to submit annual financial statements, and set down prosecution procedures and penalties.[10] Below is a sample freight-rate schedule taken from the law:

---

8 *Ibid.*, 34–35.
9 *Ibid.*, 46; Mildred Throne, "The Repeal of the Iowa Granger Law, 1878," *Iowa Journal of History*, LI (April, 1953), 98.
10 Iowa *Laws* (1874), 61–89.

Distance in miles	Wheat in cts. per 100 lbs. per car load	Cattle & Hogs in $ per carload	Class-A freight, in $ per carload
50	9.43	$16.28	$22.63
100	14.26	21.38	30.90
200	18.67	31.37	41.20
300	22.35	39.86	49.75
350	23.80	43.86	53.22[11]

Contrary to developments in Iowa, alarm did not sweep through the river towns of Illinois, Minnesota, and Wisconsin when railroads arrived. Agitation for controls in these states came from interior waypoint interests and elsewhere. In fact, the legislation provided by the Wisconsin General Assembly "was little more than a Republican trick designed to embarrass a Reform party administration." [12]

The Grange's role in proceedings which produced railway controls can best be determined by looking at state records. For Illinois, archives at the state capitol present the most accurate picture of the order's relationship to the regulatory acts passed by the legislature. Late in the 1860s when the general assembly first weighed possibilities of controlling railroads, the Grange still was in its infancy. Consequently, it is not surprising that the Illinois State Archives' file-cabinet drawers marked "Petitions Regarding Railroad Rates" and "Misc[ellaneous] Petitions" contain none from granges in that decade. Apparently, the few chapters which existed then mounted no campaigns and collected no petitions for distribution to state lawmakers. More than likely, electing officers, drafting bylaws, learning ritual, proselytizing, and erecting grange halls were the first considerations of the early grangers; legislative battles came later. In fact, it was not until after the so-called granger laws had been enacted that the voice of the order was heard on political matters. Thus, the act approved March 10, 1869, for the control of common carriers in Illinois had no connection with the farmers' brotherhood. The 1869 law required railroads to post their passenger and freight tariffs, prohibited rebates and drawbacks, and fixed penalties. Although the enactment did not provide for the establishment of specific rates, it represented a beginning.[13]

11 *Ibid.*, 62–75.   12 Miller, "Origins of the Iowa Granger Law," 677–80.
13 Illinois General Assembly, Petitions Regarding Railroad Rates, 1869–74, in Illinois State Archives, Springfield; Illinois *Laws* (1869), 309–11. Consult as well: Harold D. Woodman, "Chicago Businessmen and the 'Granger' Laws," *Agricultural History*, XXXVI (January, 1962), 16–24.

Generally, the forces pushing for railroad regulations in Illinois during the sixties spoke for no special interest groups and petitioners came from no particular section of the state. Petitions were received from such diverse places as Edwards County in southeastern Illinois and Carroll County in the northwestern tip of the state. From the evidence, only one safe conclusion can be reached concerning the activity. Apparently, long-and-short-haul discrimination hurt the small townsman most severely. Almost every petition was sent by an independent citizenry group from an isolated downstate county.[14]

If the forces which produced legislation in 1869 had been strong enough to obtain regulation then, there is no reason to suspect that their power declined in the seventies. Moreover, there is nothing to indicate that their desire for effective regulations was satisfied by the 1869 law. To assume that the Grange emerged to take up the slack caused by the appeasement of nonagricultural interests in 1869 is presumptuous. In fact, petitions in the state archives suggest the opposite conclusion. If anything, desires for tight railroad controls among nonfarmers increased after the passage of the 1869 act. A total of 110 petitions asking for more regulation reached the legislature in 1873 and 1874, but only two of these came from grangers. Interestingly enough, independent farmers' clubs showed more concern for the status of common carriers in Illinois than did chapters of the Patrons of Husbandry. Nevertheless, nonagricultural interests were more involved in the struggle to correct transportation imbalances than were farmers.[15]

Inundated with requests seemingly from all quarters, the Illinois General Assembly responded twice during the early seventies to the demands. On July 1, 1871, a bill to establish a railroad and warehouse commission became law. Among other things, it set standards for board membership and empowered commissioners to examine railway company books, question witnesses, conduct hearings, and penalize offenders. When the 1871 law proved inadequate, the legislature resumed the debate on railroad regulations at the 1873 session. The atmosphere soon became charged with antirailroad and prorailroad biases, and passions became heated on both sides. In the end, however, foes of railroads triumphed as the reformers in the general assembly pushed through the most comprehensive railroad control bill ever passed by an Illinois state legislature.

14 Illinois General Assembly, Petitions Regarding Railroad Rates, January 17, 1865, February 1, 1869.
15 *Ibid.*, 1873–74.

There were eleven parts to the Illinois railroad law of 1873. The first two sections declared that any railroad line charging more than a fair or reasonable rate would be deemed guilty of extortion, and any carrier discriminating unjustly would be in violation of the law. Section three defined in precise terms what was meant by discrimination. The next two sections fixed the fines to be collected from offenders. Section six provided the aggrieved party with the right to sue for the recovery of three times the amount of damages, plus court costs and attorney's fees. The seventh section charged the railroad commissioners with personally investigating and ascertaining whether the articles of the act had been violated and with immediately causing "suits to be commenced and prosecuted against any railroad corporation which may violate the provisions of this act." By the eighth section of the law, the commission was directed to prepare a maximum-rate schedule for the transportation of passengers and cargoes by each railroad operating in Illinois. These rates would be taken after January 15, 1874, as *prima facie* evidence of reasonableness in all cases involving charges against carriers. Such schedules were to be reviewed periodically by the commission as circumstances required. Section nine provided that "in all cases under the provisions of this act, the rules of evidence . . . [should] be the same as in other civil actions, except as hereinbefore otherwise provided." In the next section, the "term 'railroad corporation,' " as used in this law, was "taken to mean all corporations, companies, or individuals now owning or operating, or which may hereafter own or operate any railroad . . . in this state." The final section of the law was a simple declaration calling attention to the fact that the 1873 act was replacing the railroad law of 1871.[16]

Support for the new law came from a variety of sources. For the Illinois campaign to regulate common carriers, the old saying that "politics make strange bedfellows" should be reworded to read "demands for railroad regulations make strange bedfellows" because such diverse interests as the Grange and the Chicago Board of Trade worked for and praised the 1873 law. For overall influence, however, the latter organization was the more decisive force. Lee Benson's conclusion is essentially correct: "the Illinois Board of Railroad and Warehouse Commissioners was not so much the product of spontaneous indignation on the prairies as a monument to the strategic talents of the Chicago Board of Trade." [17]

16 Illinois *Laws* (1871), 618–25; (1873), 135–40.
17 Chicago *Prairie Farmer*, June 21, 1873; Illinois General Assembly, Petitions Re-

Actions taken in Minnesota and Wisconsin to control common carriers closely resemble developments in Illinois. In the former state, Rasmus S. Saby has pointed out that concern with the practices of transportation companies predated the formation of the Grange. Between 1862 and 1870, in amending those charters enacted during the territorial days of Minnesota for the control of common carriers, the general assembly frequently inserted the provision that common carriers were to carry all passengers and freight delivered to them by connecting lines on the same terms and at the same rates that were given originated traffic. Moreover, several special railroad laws passed during this period bound railroad companies to transport cargo and passengers at "reasonable" rates.[18]

In 1867 the congressional land grant of the previous year was given to the Southern Minnesota Railroad Company, "provided, that the legislature shall have the right to fix and regulate from time to time the rates of freight and passenger tariffs on said railway, or on any branch or division thereof." Debates soon arose in the general assembly as to whether the legislature had constitutional power to govern tariffs and whether the exercise of such authority conflicted with judicial responsibilities. These questions were then referred to the attorney general, and he quickly closed the subject by concluding that rate fixing and regulating were not the bailiwick of the legislature. As a footnote, he added that rates so fixed were not "of binding force or validity." [19]

Pressures mounted in Minnesota for effective railroad legislation during the early seventies, so the state legislature enacted four laws. The acts represented a progression of tighter and tighter controls. The first of these, the Commissioner Act of 1871, established a state regulatory agency and defined the office's responsibilities. The law directed commissioners to examine the operations of Minnesota rail lines, and it called on these officials to prepare annual reports on the conditions of transportation systems in the state. A companion bill enacted the same year fixed maximum railroad tariffs for lines operat-

---

garding Railroad Rates, March 4, 1874. See as well Woodman, "Chicago Businessmen and the 'Granger' Laws," 16–24; Lee Benson, *Merchants, Farmers, and Railroads; Railroad Regulation and New York Politics, 1850–1887* (Cambridge, 1955), 25.

[18] Rasmus S. Saby, "Railroad Legislation in Minnesota, 1849–1875," *Minnesota Historical Collections*, XV (May, 1915), 62–63.

[19] *Ibid.*, 63–64.

ing within Minnesota, forbade discrimination, and laid down a variety of less important guidelines. The law provided for the classification of freight and permitted rate variations based on mileage as long as the tariffs did not exceed the maximum levels permitted by the law. For example, the highest tariff which common carriers were permitted to charge for transporting foodstuffs fewer that twenty-one miles was fixed at six cents a ton per mile; the top rate for carrying the same goods twenty-one to fifty miles was set at five cents a ton per mile. In contrast, a flat maximum charge of five cents per mile was established for passenger travel.[20]

In 1874 the general assembly repealed the acts of 1871 and replaced them with a more comprehensive law. The omnibus bill enacted March 6, 1874, provided for a three-man board of railroad commissioners and placed great powers at their disposal. Board members were empowered to investigate railroads, examine company books, subpoena witnesses, prosecute offenders, and fix rates. Moreover, the law carefully defined what was meant by discrimination and laid down penalties for specific offenses. By a supplementary act approved March 9, 1874, the legislature established standards for warehouses and elevators owned by railroad companies. The law stated that their fees for "receiving, elevating, handling, and delivering" grain were not to exceed two cents per bushel.[21]

In Wisconsin, the effect of the Grange on the movement which produced railroad regulation was more in evidence than it was in any of the other states. And yet, as Herman J. Deutsch has pointed out, the role of the farmers' order was insignificant when it is compared to other forces. Railroad controls in Wisconsin resulted more from political expediency on the part of the GOP than from anything else. Republicans knew they had to seize the issue of railroads to survive as the dominant force in Wisconsin politics, so they pushed through a railroad-control bill which was so strong that it left third-party forces without an issue. Thus the GOP used the farmers' old nemesis to accomplish one of its chief objectives in the early 1870s; it had outflanked the reformers to such degree that the Republican party had only to worry about Democrats in future elections. In the minds of many politicians grangers were responsible for the political unrest in Wisconsin. That view exaggerated the importance of farmers, but Republicans still felt that the enactment of railroad regulation would

20 Minnesota *Laws* (1871), 56–59, 61–66.   21 *Ibid.* (1874), 140–50, 155–57.

quiet reformers and remove the wind from dissidents' sails. For these reasons the legislature approved the Potter bill for railroad control.[22]

The Potter Railroad Law of 1874 had eighteen parts. The first four sections pertained to classification of Wisconsin railroads and to criteria used to determine rates for each class. Here was the heart of the law; that elaborate schedule of tariffs and fares was the most important feature of the act. The next three sections outlined procedures to be employed in prosecuting alleged violators and fixed penalties to be meted out against lines found guilty. Articles 8 through 17 provided for the creation of a railroad commission with specifically prescribed duties relating to the enforcement of the act. Finally, the law stated in the eighteenth section that its provisions applied only to intrastate shipments.[23]

Grangers worked in close contact with Governor W. R. Taylor to see that able men were selected for the railroad commission. From the beginning, J. H. Osborn of Oshkosh was most acceptable to grangers as the head of the board. His name was first proffered March 17, 1874, at a meeting called by the governor. Evidently, Taylor agreed with the order's choice because Osborn received the appointment to the position.[24]

With Osborn firmly entrenched in the commissioner's chair and with an adequate railroad law on the statute books, all that was left for the order to accomplish in relation to common carriers was legislation for the repeal of special tax exemptions given railroads. By their charters, many roads had received favored status from the state, thereby enabling their properties to go untaxed. Grangers viewing this state of affairs saw handsome profits going to railroads, so they petitioned the legislatures for changes. Their pleas produced nothing in the way of legislation.[25]

The story in Illinois was similar. Grangers resolved that railroads

22 Herman J. Deutsch, "Disintegrating Forces in Wisconsin Politics of the Early Seventies," *Wisconsin Magazine of History*, XV (June, 1932), 393–94; Wisconsin State Grange, *Proceedings*, January, 1874, 23. Dale E. Treleven has shown the importance of Francis H. West and other grain commissioners from Milwaukee in the struggle for enactment of railroad regulations. Examine: Treleven, "Railroads, Elevators, and Grain Dealers: The Genesis of Antimonopolism in Milwaukee," *Wisconsin Magazine of History*, LII (Spring, 1969), 209–11, 215. For a discussion of the Grange and political partisanship in Wisconsin, see the preceding chapter.

23 Wisconsin *Laws* (1874), 599–606.

24 Wisconsin *General Assembly Journal* (1875), 167; Wisconsin State Grange, *Proceedings*, 1875, pp. 42–43; Appendix, 3–6.

25 Wisconsin State Grange, *Proceedings*, January, 1874, p. 36; 1875, p. 61; Wisconsin *General Assembly Journal* (1875), 174.

deserved no special tax considerations before the law, but the state legislature viewed the matter differently. Nothing was done immediately to remove the exemptions granted railway lines by their charters.[26]

In the four midwestern states where railroad regulations had been enacted, court cases arose almost immediately to test the statutes. Known popularly as the "Granger Cases," these suits stand as landmarks in the judicial proceedings of the United States because of the decisions rendered by the Supreme Court. In effect, the majority decided that "where property has been clothed with a public interest, the legislature may fix a limit to that which shall in law be reasonable for its use." In one form or another, this principle was found in all of the eight cases decided by the Supreme Court in March, 1877. With the Supreme Court on the side of regulation, railroads had one of two courses of action open to them. They could accept controls and hope for the best, or they could oppose measures and fight for their repeal. Whenever lines pursued the latter alternative, grangers defended railroad regulations and petitioned in their behalf. They simply did not want to return to the era when the shipper and the traveler were at the mercy of capitalists. Occasionally, as in the case of Iowa, their efforts failed; legislators there found it to their advantage to listen more attentively to the pleas of railroaders than to the wishes of farmers and others dependent upon cheap and equitable rail transportation.[27]

Railroad offenses were just as abusive in the West, East, and South as they were in the upper Mississippi River Valley, and the granger response was essentially the same everywhere. Since the farmer in Kansas confronted the same transportation problems as his counterpart in other parts of the nation, his demands for legislation had a familiar ring. Consequently, grangers throughout the country fought discrimination, high tariffs, and free passes to lawmakers, and they

26 Champaign County Pomona Grange, Petition Dated January 19, 1877, from the La Salle County Pomona Grange No. 61, Champaign County Pomona Grange Record Books.

27 Illinois Railroad and Warehouse Commission, *Seventh Annual Report* (1877), Appendix, 5–6, 20, 22, 24–25; Elwin W. Sigmund, "The Granger Cases: 1877 or 1876?," *American Historical Review*, LVIII (April, 1953), 571–74. Wisconsin *General Assembly Journal* (1878), 147; Throne, "The Repeal of the Iowa Granger Law," 97–130. The eight cases were: *Munn v. Illinois*; *Peik v. Chicago and Northwestern Railway Company*; *Lawrence v. same*; *Chicago, Burlington and Quincy Railroad Company v. Iowa*; *Stone v. Wisconsin*; *Winona and St. Peter Railroad v. Blake*; *Chicago, Milwaukee and St. Paul Railroad Company v. Ackley*; *Southern Minnesota Railroad Company v. Coleman*.

asked for the establishment of state railroad commissions with powers to oversee the activities of railway companies and to enforce laws against unfair practices.[28]

Many myths concerning railroad regulation still exist despite the efforts of Benson, Saby, and Miller to expose them. Many textbooks leave the impression that railroad regulation originated in the upper Mississippi River Valley; in fact, nothing could be further from the truth. Rhode Island pioneered in railroad legislation. In 1839 the legislature of that state set an example which was followed first by the states of the East and then by the states of the upper Mississippi River Valley. The general assembly of Rhode Island vested an individual with the title of railroad commissioner and authorized him to oversee the state's railroad operations.[29]

Also missing from most studies of the Grange is mention of the railroad legislation which controlled the intrastate operations of common carriers outside the Upper Mississippi River Valley. For some mysterious reason, the laws governing railroads in Illinois, Iowa, Minnesota, and Wisconsin have always been known as the granger laws; similar measures enacted elsewhere during the same era have not been so designated. Generally, it is not even known that railroad-control measures were considered and passed by the general assemblies of Ohio, Maryland, Delaware, Michigan, and Indiana during the same years that the lawmakers of the so-called granger states were acting on proposals for similar legislation. Admittedly, legislation passed outside the Valley was not as comprehensive as the laws enacted by the general assemblies of Illinois, Iowa, Minnesota, and Wisconsin. But as Professor Lee Benson notes, "Commercial groups throughout the country, particularly those representing major coastal and interior cities, dominated the campaigns centering on transportation issues." Therefore, if anything produced "the first major phase of postwar railroad agitation," it was place discrimination that turned businessmen against common carriers and forced them to seek controls.[30]

[28] Macon *Missouri Granger*, October 19, 1875; New York *Senate Journal* (1881), 150; Ohio State Grange, *Proceedings*, 1891, pp. 28–29; 1897, pp. 46–48; Maryland State Grange, *Proceedings*, 1891, p. 15; Kansas *House Journal* (1874), 524–25; Kansas State Grange, *Proceedings*, 1879, p. 9; South Carolina State Grange, *Proceedings*, 1879, n.p.; North Carolina State Grange, *Proceedings*, 1878, p. 17; Oregon State Grange, *Proceedings*, 1887, p. 40; Illinois State Grange, *Proceedings*, 1891, pp. 57–58; Richmond *Virginia Patron*, February 16, 1877.

[29] Saby, "Railroad Legislation in Minnesota," 104–109.

[30] Miller, "The Granger Laws," 65–66; Michigan *Laws* (1869), I, 182; Benson, *Merchants, Farmers, and Railroads*, 23.

To suggest that all grangers placed faith in regulations to solve the nation's transportation problems would be exaggerating reality. Fiery Ignatius Donnelly was probably the leading member of the order who looked for no miracles from regulatory measures. Judging by what he said and advocated, he had little or no confidence in legislative controls per se. His suspicions of business and his fear of the power of money were much greater than his trust in republican institutions. Fear especially played a great part in Donnelly's reservations about the effectiveness of controls. He, like many other members of the organization, was afraid that railroads possessed the will and the means to "buy up" legislatures and courts. Therefore, it is not surprising to find Donnelly and others in the order urging grangers to seek solutions outside the realm of controls. At one point, Donnelly called on the government to construct a canal between St. Paul and Duluth. With such an artery available to Minnesota farmers, he hoped that agriculture's dependency upon railroads would be diminished. A waterway between these points would afford farmers an opportunity to ship wheat directly to Europe by way of the canal, the Great Lakes, the St. Lawrence River, and the Atlantic Ocean. Minnesota grain would find its way to Liverpool to compete favorably on the world market, and railroads would find themselves at the mercy of farmers.[31]

Grangers did not always look to the law for protection from railroads. There were instances in which order chapters used cooperation to provide better transportation services. Florida grangers owned and operated a steamboat on the St. John's River and thereby reduced cotton shipping costs by $1.50 a bale in 1875. Meanwhile, California members raised enough money by stock subscriptions to build a narrow-gauge railroad from Monterey to a point twenty miles inland. Despite the fact that the railroad and its appurtenances cost the people of Salinas Valley $13,000 a mile, proponents believed that the new line would carry 1.5 million bushels of wheat in 1874 at reduced freight rates and still yield dividends to the road's stockholders.[32]

Attesting to the influence of the Grange in the movement for railway regulations were the privileges granted members of the farmers' brotherhood by railroad companies. To curry the favor of grangers, railroad companies often granted special privileges. It was not un-

[31] St. Paul *Reporter*, April 25, 1873. A more detailed sketch of the order's role in campaigns for improving water transportation routes will follow the paragraphs on railroads.

[32] Macon *Missouri Granger*, May 18, 1875; Friendship, Wisconsin, *Adams County Press*, March 27, 1875.

common, for example, for lines to carry grange delegates to and from annual meetings at reduced fares; nor was it unusual for railroad companies to transport at special rates grangers and their stock and other articles to be exhibited at agricultural fairs. In southwestern Minnesota, the St. Paul and Sioux City Railroad even helped farmers to obtain trees and shrubs for their prairie homesteads. Master George I. Parsons of the state grange was one spokesman who was impressed with the generosity of the line. On a trip through the pioneer sections of Minnesota in 1873, he recalled noticing "all along, cars laden with trees and cuttings, to be planted on those vast prairies, and was informed, by posters in all the depots and stations, that the transportation of such trees and cuttings would be free to all who desire to plant them." Parsons noted in an article for the *Farmers' Union* that "this liberality, on the part of the companies, is worthy of all praise, and is winning golden opinions from the pioneers." Therefore, because of the favors received and the gratitude expressed, it would not be wise to lose sight of the fact that grangers appreciated what railroads were doing for them; all they wanted as shippers and as passengers was fair treatment. An objective observer must agree with the Tennessee granger who said that the brotherhood had "no war to make on railroads" because farmers could not "afford to do without them"; the only thing the order wanted was "fair dealing." [33]

When contrasted to order demands for state railroad regulations, granger participation in movements for federal controls of interstate commerce seemingly followed a pattern of increasing involvement. During the first ten years of the order's existence, the need for federal regulations was not explored very often at grange sessions. Among the officers, National Master Dudley W. Adams seemed to be the only one who was demanding federal action. He mentioned possible benefits of federal intervention in his annual addresses of 1874 and 1875, but nothing ever came of his suggestions. Apparently, the National Grange was not ready to see the federal government's responsibilities broadened to the extent that Washington would have jurisdiction over the affairs of private businesses.[34]

[33] Denver *Rocky Mountain Weekly News*, March 18, 1874; Maine State Grange, *Proceedings*, 1874, p. 17; Massachusetts State Grange, *Proceedings*, 1884, p. 64; New Hampshire State Grange, *Proceedings*, 1890, pp. 117–18; New Hampshire Grange Fair Association, *Official Premium List*, 8; Minneapolis *Farmers' Union*, May 3, 1873; Indiana State Grange, *Proceedings*, 1874, p. 30; Illinois State Grange, *Proceedings*, 1892, p. 63; Tennessee State Grange, *Proceedings*, 1875, p. 31.
[34] National Grange, *Proceedings*, 1874, pp. 14–16; February, 1875, p. 11.

The disinterest of the Grange in the early seventies in the question of federal regulation appeared in the records of the Senate Committee on Transportation Routes to the Seaboard. Chairman William Windom invited no grangers to testify before his committee, but President W. C. Flagg of the Illinois State Farmers' Association was afforded an opportunity to appear. The only feasible explanation for the conspicuous absence of order spokesmen at these hearings revolved around the fact that the organization had not yet taken a stand in 1873 when the testimony was being collected. Consequently, the committee's star witnesses proved to be either representatives of transportation concerns, members of boards of trade, leaders of independent farmers' clubs, or engineers.[35]

Granger activity in behalf of national railroad regulation became more evident in the late seventies, and apparently the sudden flurry of interest stemmed from the decisions rendered by the Supreme Court in 1877. By its judgments in the "granger cases," the court inadvertently caused a shift of direction. Shippers who had looked to state capitols for protection now were forced to agree with the justices who said the states had no constitutional jurisdiction over interstate commerce. For the Grange, the court's verdicts meant that the order would have to move its campaign for more equitable railway service to Washington if it wanted to obtain legislative solutions which would meet the specifications laid down by the Supreme Court. Subsequently, more and more grangers acquainted themselves with the facts and their organization's drive for federal railway legislation picked up additional momentum with each successive year. Results of the added emphasis on federal solutions were in evidence everywhere. Reference was made to the need for federal regulations in dozens of reports, resolutions, and petitions.[36] At the same time, granger goals in regard to federal regulations were being refined and crystallized. By the mid-eighties, leading grangers were announcing that the members of their brotherhood would not be placated if Congress provided them with inadequate legislation. Moreover, they were now favoring the Reagan bill over the Cullom bill because the former proposal

[35] U.S. Senate, Report of the Select Committee on Transportation Routes to the Seaboard, *Senate Report* 307 (Washington, 1874).

[36] Illinois Railroad and Warehouse Commission, *Seventh Annual Report*, Appendix, 5–6, 20, 22, 24–25; Virginia State Grange, *Proceedings*, 1879, p. 6; Oregon State Grange, *Proceedings*, 1881, p. 13; National Grange, *Proceedings*, 1882, pp. 58–60; 1886, pp. 124–30; Texas State Grange, *Proceedings*, 1882, pp. 10–11; Denver *Colorado Farmer*, January 22, 1885; Vermont State Grange, *Proceedings*, 1886, pp. 20–22.

was more comprehensive and had fewer loopholes than its rival.[37]

Master Leonard Rhone of the Pennsylvania State Grange expressed the Grange's position as well as anyone in the brotherhood. In a guest lecture presented to the Vermont State Grange in 1886, Rhone explained why the coverage and safeguards provided by the Reagan bill were superior to those offered in the Cullom proposal:

> The two bills differ radically, and the fact should be borne in mind in considering their relative merits that the Cullom bill has been earnestly advocated by the attorneys of railroad companies, while the Reagan bill has been opposed by them. The Cullom bill depends upon a Commission of nine men for its enforcement through existing Courts. The Reagan bill presented the three great essentials in legislation upon this subject: 1st, Prohibition of greater charges for the short distance than for longer ones which includes the less; 2d, full publicity of rates, and strict adherence to them by railroads; and 3d, absolute prohibition of pooling. The Reagan bill lays down general principles which are uniform in their application, while the Cullom bill provided that the short haul principle and publicity of rates may be suspended at the pleasure of the Commission.[38]

Despite the fact that the Grange had made its will known to Congress on several occasions through petitions and memorials, support for the order's cause was not strong enough to overcome the forces leaning in the direction of the Cullom bill. As a result, grangers had to settle for the latter bill.[39]

Therefore the Grange deserves little credit for the enactment of the Interstate Commerce Act. Solon Buck summed up the situation well when he stated that the "claims of certain Grange enthusiasts that the credit for this act belongs to the order of Patrons of Husbandry can hardly be substantiated for an examination of the report of the Cullom committee shows that the demand for regulation of railroads was fully as insistent among merchants and manufacturers at this time as among farmers." Gabriel Kolko and Gerald D. Nash have shown why there is little historical basis for the claims made by the order's propagandists concerning granger responsibility for the passage of the Cullom Act. To the latter historian, the independent oil producers and refiners of Pennsylvania "must ... be counted among the most important groups pressing for government regulation, for it was their direct influence which in 1878 led to the introduction of the Reagan bill in the House of Representatives." Nash went on to explain how

[37] National Grange, *Proceedings*, 1882, pp. 58–60; 1886, pp. 124–30.
[38] Vermont State Grange, *Proceedings*, 1886, pp. 20–22.
[39] *Congressional Record*, 50th Cong., 1st Sess., Pt. I, 183–84.

will to survive had forced the independent oilmen of western Pennsylvania to ask Congress early in 1876 for the regulation of railroads.
The threat posed by the creation of John D. Rockefeller's Central Association for the gaining of rebates to the Standard Oil Company by
the New York Central, Erie, and Pennsylvania railroads had caused
the independents to seek relief.[40]

Prodded by Representative James H. Hopkins of Pittsburgh, Congress held hearings to investigate the oilmen's charges. The findings
proved inconclusive. Apparently, the investigators produced nothing
substantial because they lacked the resolve necessary to uncover the
truth. Undaunted, Hopkins introduced a bill prohibiting rebates and
discrimination and requiring freight carriers to post their rate schedules. In the course of his fight, the lawmaker from the Keystone State
enlisted the support of John H. Reagan of Texas. Nevertheless, despite wider backing, the Hopkins bill of 1876 still died in committee.
Subsequently, the oil war intensified. As it did, independents united
for one last effort against the power of Standard Oil. During the
course of the battle, Reagan, supported by the oilmen and by the 1877
decisions of the Supreme Court, reintroduced the Hopkins bill in
1878. This time the measure passed the House but died in the Senate.
Feeling that their last hopes for survival had been dashed by the defeat, independents surrendered peacefully to Rockefeller's mammoth
company. By the fall of 1880, all that remained of their efforts was a
legacy of interest in the regulation of interstate railroads.[41]

Meanwhile, Gabriel Kolko concentrates on the role railroads
played in movements for interstate commerce controls. He surmises
that the railroad companies themselves, and not farmers and shippers,
"were the most important single advocates of federal regulation from
1877 to 1916." Presumably rail transportation companies had tried
and failed in their attempts to govern their own industry. Their failures soon brought railroaders to the realization that federal intervention would be more acceptable than another round of cutthroat
competition. In the course of debates over control of carriers, many
railroad men admitted that they would not oppose the idea af government rate fixing, "the alleged bogey," if it were accompanied by
legalized pooling.[42]

---

[40] Buck, *The Granger Movement*, 230; Gerald D. Nash, "Origins of the Interstate
Commerce Act of 1887," *Pennsylvania History*, XXIV (July, 1957), 181–90.

[41] *Ibid.*

[42] Gabriel Kolko, *Railroads and Regulation, 1877–1916* (Princeton, 1965), 3–6,
28–29.

For many grangers, enactment of the Cullom bill in 1887 was not an end; it was only a beginning. Consequently, they paid close attention to railroads operating across state lines in order to detect violations and to discover loopholes in the law. On at least three occasions between 1887 and 1892, members reported to the Interstate Commerce Commission that the law was not functioning to their satisfaction. In 1887 the Vermont State Grange charged the Central Railroad Company with not living up to the intent of the law. Specifically, the New Englanders were disgruntled because they were certain that the line practiced long-and-short-haul discrimination in violation of the fourth section of the Cullom Law. Upon investigating the complaint, the ICC found the line guilty and ordered it to reduce its tariffs by a third.[43]

Grange committees on transportation in Delaware and New Hampshire also found instances of place discrimination occurring after enactment of the Interstate Commerce Act. Transport costs were greater from northern New Hampshire to Boston than they were from Chicago to the Massachusetts city, and this disturbed grangers in the Granite State as much as rate discrimination angered members in Delaware. Farmers there certainly found themselves in a disadvantageous position. They paid more to ship freight to Jersey City than was charged Norfolk shippers to dispatch similar quantities to New York City. At the same time, a cargo of Georgia peaches found its way to New York at a rate which was comparable to that assessed Delaware orchard growers to ship fruit to the same point. To make matters worse, Boston-to-Chesapeake Bay shipments were no cheaper than coast-to-coast movements.[44]

Since the ICC had failed many times to produce desired changes, grangers turned increasingly to more radical solutions. Several chapters called on the federal government to take over the operation of the nation's railroads, but overall the movement for nationalization never gained very wide acceptance in the Grange. Unlike Populists and Knights of Labor, most members of the oldest national farmers' brotherhood were too conservative to ask for such drastic steps.[45]

Grange interest in railroads often went beyond obvious topics of

[43] Greenville (S.C.) *Cotton Plant*, December, 1887; New Hampshire State Grange, *Proceedings*, 1892, p. 80.

[44] Delaware State Grange, *Proceedings*, 1892, pp. 44–46.

[45] Colorado State Grange, *Proceedings*, 1888, p. 18; Wisconsin State Grange, *Proceedings*, 1896, p. 38; National Grange, *Proceedings*, 1880, pp. 74–78; Illinois State Grange, *Proceedings*, 1897, pp. 32–33; 1899, pp. 45–46.

concern. The degree to which a community was isolated usually determined whether bestowing land grants to railroads was a matter of controversy. Granges located near major arteries of transportation viewed grants to private companies as extensions of special privilege. In contrast, members from areas not served by common carriers were less philosophical and more pragmatic. For them, obtainment of transportation lines was the only question at stake. If governmental aid brought railroad service to them, they were for it. The National Grange took sides in this controversy only once. In 1875 it made the mistake of backing residents of the Southwest who favored the granting of governmental land to the Southern Pacific. In the opinion of the more zealous antirailroad men in the order, the parent body's action represented treason which must be "condemn[ed] ... in the strongest terms." [46]

Rounding out the Grange's transportation program were demands for better roads and improved waterways. Over the span of three decades, chapters insisted that canals be dug to link the St. Lawrence and the Richelieu River in Vermont, the Wabash River and Lake Michigan in Indiana, the Great Lakes and the Illinois River in Illinois, and the Fox, Wisconsin, and Mississippi rivers in Wisconsin. Moreover, during the same period, grangers were expressing the need for navigational improvements. The Columbia, Willamette, Snake, Cumberland, Tennessee, and Mississippi rivers were among the many waterways singled out for ameliorations. According to the petitioners, these rivers either needed to have their channels deepened or their obstructions removed.[47]

Among the more knowledgeable grangers, there was a recognition that increased foreign trade offered American farmers the greatest possibilities for relief from continually falling prices. At the same time, there was also an awareness of the factors which accounted for high shipping costs at several ports. Familiarity with harbor and wharf facilities along the three coasts gave many grangers knowledge of the bottlenecks which existed at numerous ports. Moreover, the

[46] Indianapolis *Indiana Farmer*, February 7, 1874; Tennessee State Grange, *Proceedings*, 1875, p. 18; Oregon State Grange, *Proceedings*, 1876, p. 22; Chicago *Prairie Farmer*, April 10, 1875; Chicago *Prairie Farmer*, April 10, 1875.

[47] Vermont State Grange, *Proceedings*, 1874, p. 14; Oregon State Grange, *Proceedings*, 1873, pp. 19–20; 1875, pp. 50–51; Indiana State Grange, *Proceedings*, 1874, pp. 30–31; Georgia State Grange, *Proceedings*, 1875, p. 46; Tennessee State Grange, *Proceedings*, 1875, p. 46; Wisconsin State Grange, *Proceedings*, January, 1874, pp. 29–33; 1875, p. 59; 1877, pp. 29–31; Illinois State Grange, *Proceedings*, 1889, p. 57; 1899, p. 34; *Congressional Record*, 50th Cong., 1st Sess., Pt. I, 183–84.

need for an Isthmian canal was obvious to anyone who had waited for a transcontinental shipment to arrive. Therefore, the Grange tied the need for more international commerce to its requests for the outfitting of a canal across Nicaragua and for the improvement of harbors, wharves, and shipping services in general. In 1874 the executive committee of the Missouri State Grange even went so far as to suggest that direct trade with Brazil might be contemplated if the mouth of the Mississippi River were improved. In a circular distributed among Mississippi Valley grangers, plans were spelled out and possible advantages accruing from such trade were listed. Although the proposal undoubtedly stimulated thought, no important movement developed to guide the plan through Congress to a successful conclusion.[48]

Grangers generally expected to see better transportation services offered at lower prices, but their actions were not always consistent with their aspirations. One of the most interesting ironies involved the reluctance of members to associate formally with the American Cheap Transportation Association (ACTA) in the early seventies. As its name implies, the association was a league of individuals bent on obtaining cheaper transportation rates. Although ACTA membership cut across class lines, its leadership had a definite aristocratic makeup. Wealthy landowners, large planters, and representatives of commercial boards dominated its proceedings from the outset. On May 7, 1874, Francis B. Thurber, a partner in one of New York City's largest wholesale grocery houses, emerged from two days of organizational meetings at the old Astor House in New York to be the body's "real driving force." Although the ACTA had the sympathy of the Grange, it never had the order's full support. Prominent grangers attended and participated in the ACTA meetings in New York and Washington, but little came from these contacts. Apparently, members of the farmers' brotherhood did not trust the motives of the ACTA leadership, so they did not align themselves with the national organization that had as its goal cheaper transportation for everyone.[49]

Interested as grangers were in improved transportation facilities

---

[48] National Grange, *Proceedings*, February, 1874, pp. 78–79; 1894, p. 196; 1895, p. 178; Oregon State Grange, *Proceedings*, 1881, p. 14; Maryland *Senate Journal* (1876), 99, 108; Mississippi State Grange, *Proceedings*, 1875, p. 15; California State Grange, *Proceedings*, 1891, pp. 71–72; Circular entitled *Direct Trade with Brazil*, 1874, issued by the Executive Committee of the Missouri State Grange, in Osborn Papers.
[49] Benson, *Merchants, Farmers, and Railroads*, 59–60; Wisconsin State Grange, *Proceedings*, 1874, p. 36; New Hampshire State Grange, *Proceedings*, 1874, p. 36.

and services, it is surprising to find that their recognition of the need
for good roads did not develop until the nineties. For many years,
they were so absorbed in rail and water transportation problems that
they lost sight of the need for adequate roads. Then, with the enactment of the Cullom bill and the advent of the bicycle craze, attention turned suddenly to roadways. It seemed then as if everyone in
the organization was talking about the value of highways to farmers.
They specifically pointed out that good roads aided tourism, made
"rural life more attractive," and benefited trade and communications
generally.[50]

For some grangers, the provision of highways was a county matter.
Overwhelmingly, however, the bulk of the membership considered
road construction a concern best handled by the states. Only a few
members saw a place for the federal government in this activity. Bicycle enthusiasts and manufacturers also championed the cause of good
roads, but they did not necessarily agree with the proposals put forward by granges. In Ohio, for example, the cycle industry announced
in 1893 that it favored the floating of government bonds to pay for the
macadamization of every road in the state. In response to this suggestion for debt financing, the state grange said that it could not accept
any scheme for paving roadways unless it was based on "pay as you
go." To the cyclists of New Hampshire, the state grange must have
appeared especially cantankerous because of its suggestion in 1898 for
a tax on bicycles to pay for highway maintenance.[51]

As grangers looked more intensely at road problems, they saw other
issues that had previously eluded them. For one thing, they developed
an awareness of the need for maintaining safe locomotive crossings.
Beginnings were also made in the field of tire control. In 1899 the
Illinois State Grange had under study a proposal for the establishment of state-enforced, tire-width specifications, but the chapter ta-
bled the resolution. Earlier, it had received an unfavorable report
from the committee on taxation and legislation because the committee members unfortunately had no comprehension of the relationship between road wear and wagon wheel sizes. The fact that

50 New Hampshire State Grange, *Proceedings*, 1891, pp. 110–12; 1892, pp. 14–15,
19–20; Church Hill Grange, Secretary's Minutes, February 22, 1878; Social Grange,
Secretary's Minutes, December 6, 1891.
51 Church Hill Grange, Secretary's Minutes, February 22, 1878; New Hampshire
State Grange, *Proceedings*, 1891, pp. 110–12; 1898, p. 142; Illinois State Grange,
Proceedings, 1892, pp. 12, 54; 1893, p. 13; New York State Grange, *Proceedings*,
1894, pp. 104–105; Ohio State Grange, *Proceedings*, 1893, p. 8.

some members did showed the degree to which some grangers were thinking.[52]

In the field of communications, grangers were interested in obtaining better mail, telegraph, and telephone service. Postal reforms were certainly needed. Most Americans were still receiving their mail at village post offices as late as 1890. Home deliveries did not begin in New York City until 1863, and it was not until 1887 that towns of ten thousand or more residents gained this service. More disturbing yet was the fact that most politicians and public servants refused to see the need for changes. They no doubt feared the costs involved in improving the system. Postmaster General John Wanamaker, therefore, deserves the title of "father of the modern postal system" because he was the one individual who defied tradition more than anyone else. Most of Wanamaker's predecessors had used their positions to dispense patronage, but Wanamaker soon proved that he was going to be different. Instead of making friends for the Harrison administration, he made enemies. He alarmed bankers by advocating postal savings banks, angered express companies by demanding a parcel post system, frightened Louisiana gamblers by depriving them of use of the mails for the conduct of lotteries, enraged telegraph and telephone companies by proposing that their services be nationalized and run by the Post Office Department, and startled other conservatives by proposing the inauguration of rural home deliveries of mail at no cost to farmers.[53]

To the business community, Wanamaker was a dangerous fanatic with socialistic leanings. But to grangers, he was a friend and a defender of the rights of man. At one time or another, each of his suggestions received endorsement of the National Grange and the backing of numerous subordinate chapters. If their petitions indicate anything, they demonstrate how adamant grangers were about the desirability of receiving the last two proposals and they show how universally popular these two demands were among grangers. Literally, scores of petitions and statements of intent were drafted in behalf of rural free delivery and governmental ownership of telephone and telegraph services. Regarding the request for extensions of home de-

[52] Massachusetts State Grange, *Proceedings*, 1890, p. 80; Illinois State Grange, *Proceedings*, 1899, p. 43.

[53] Wayne E. Fuller, *RFD—The Changing Face of Rural America* (Bloomington, Ind. 1964), 13–19; Carl H. Scheele, *A Short History of the Mail Service* (Washington, 1970), 114.

liveries, the order's favorite slogan was "free rural mail delivery or no free delivery in the cities." [54]

In summary, three or four facts stand out. First, grangers were not by themselves responsible for the enactment of the so-called granger laws and the Cullom bill. Second, the importance of the Grange in the fields of transportation and communication rested more on the order's involvement than on its accomplishments. Third, grangers seldom acted in conjunction with other groups who had similar interests. Fourth, none of the major points in the order's program were unique to the degree that no other vested interest was championing essentially the same thing.

[54] Fuller, *RFD—The Changing Face of Rural America*, 13–35; Social Grange, Secretary's Minutes, December 6, 1891; Greene County Pomona Grange, Secretary's Minutes, December 6, 1894; Ohio State Grange, *Proceedings*, 1893, p. 8; 1897, pp. 46–48; Kansas State Grange, *Proceedings*, 1892, p. 8; New Hampshire State Grange, *Proceedings*, 1891, pp. 17–18; Connecticut State Grange, *Proceedings*, 1887, p. 60; Indiana State Grange, *Proceedings*, 1892, pp. 29–30; *Congressional Record*, 50th Cong., 1st Sess., Pt. I, pp. 183–84; Colorado State Grange, *Proceedings*, 1884, p. 34; National Grange, *Proceedings*, 1886, p. 134; 1888, p. 73; 1891, pp. 97–99; Wisconsin State Grange, *Proceedings*, 1877, pp. 29–31; 1896, p. 38.

Taken together, the activities of the Grange in the late nineteenth century represented a full program. The enviable record gave grangers the impression that their brotherhood deserved more recognition than it had received. To them, their order should have been credited for its role in reshaping rural America.

If the value of the Grange must be thought of in terms of one outstanding accomplishment, it would be difficult to justify any choice because the members themselves were not necessarily in agreement. To Dr. Columbus Mills, the first master of the North Carolina State Grange, "the great principle of co-operation" was established among brothers, so the order's main mission was fulfilled. To an anonymous Arkansas member whose opinions appeared April 11, 1874, in the Fort Smith *Herald*, the significant work done in reducing sectionalism was the most lasting effect of the agrarian fraternity. On the other hand, some members agreed with Master S. C. Carr of the Wisconsin State Grange, who contended that the preparation given farmers for leadership roles in society was the side of grangerism most worthy of citation.[1]

In relating what the Grange had done for him, F. G. Adams of Kansas summarized as succinctly as anyone in the order what the brotherhood had tried to do for all its members:

All know that we, in common with the farming class generally, have derived advantages from improved legislation, brought about through the influence of the grange—giving a better equalization of taxes, cheapening transportation rates, and breaking up combinations in traffic in farm products. Much has been gained through co-operation in buying and selling.

Still the more direct and appreciable advantages which are the subject

[1] Noblin, "Polk and the North Carolina Department of Agriculture," 113; Fort Smith, Arkansas, *Herald*, April 11, 1874; Wisconsin State Grange, *Proceedings*, 1886, p. 11.

of common congratulations are those which have come through the social intercourse of the grange; intercourse in the regular meetings, in the social gatherings, and in the closer neighborhood relations which grange fraternity has brought about. Through this intercourse many a grange brother and sister has become conscious of acquired power, to think and speak more clearly; we have learned to divest ourselves of narrow prejudices, and have learned lessons of charity and good fellowship in all our relations with our neighbors and with mankind.[2]

Although Adams' account was something of an overstatement, it nevertheless drew attention to the Grange's endeavors. The order sought to improve the lot of the farmer and his family by providing additional opportunities for adult education, by increasing the number of social outlets in rural America, and by developing cooperatives for economic relief. Legislatively, the Grange endeavored to obtain a better life for members of the agricultural class by petitioning in behalf of country schools and land-grant institutions, by asking for the provision of better and more equitable transportation and communication services, by demanding the enactment of various bills, and by pushing for the adoption of numerous proposals.

In retrospect, the year 1900 did not represent a break in the Grange's history. Over the years, the order and its objectives have changed very little. Testimonials to the brotherhood's consistency may be gathered from informal conversations with members of long standing in the Grange. When reminiscing they talk mostly of the enjoyment gained from gathering with neighbors for social intercourse, the monetary rewards realized by dealing through those cooperatives and mutual enterprises operated by their organization, and the instruction given during the lecture hour. Men high in the order also like to recall that grangers forced their wills on lawmakers to secure desirable legislation. The Patrons of Husbandry's first thirty-three years provided the thread which sewed the organization together and kept it a going concern through its first century of service.[3]

Therefore, with few exceptions, the "Granger's Ten Commandments" adopted in 1874 have remained almost as meaningful for members participating in the order's activities in the twentieth century as they were for grangers active in the nineteenth century.

2 Topeka *Kansas Daily State Journal,* December 10, 1885.

3 Conversations with 50-year members of Wilmington Grange No. 1918, Harveyville, Kans., October, 1966; Discussions with men close to the National Grange, December, 1966, in Washington.

1. Thou shalt love the Grange with all thy heart and with all thy soul and thou shalt love thy brother granger as thyself.

2. Thou shalt not suffer the name of the Grange to be evil spoken of, but shall severely chastise the wretch who speaks of it with contempt.

3. Remember that Saturday is Grange day. On it thou shalt set aside thy hoe and rake, and sewing machine, and wash thyself, and appear before the Master in the Grange with smiles and songs, and hearty cheer. On the fourth week thou shalt not appear empty handed, but shalt thereby bring a pair of ducks, a turkey roasted by fire, a cake baked in the oven, and pies and fruits in abundance for the Harvest Feast. So shalt thou eat and be merry, and "frights and fears" shall be remembered no more.

4. Honor thy Master, and all who sit in authority over thee, that the days of the Granges may be long in the land which Uncle Sam hath given thee.

5. Thou shalt not go to law[yers].

6. Thou shalt do no business on tick [time]. Pay as thou goest, as much as in thee lieth.

7. Thou shalt not leave thy straw but shalt surely stack it for thy cattle in the winter.

8. Thou shalt support the Granger's store for thus it becometh thee to fulfill the laws of business.

9. Thou shalt by all means have thy life insured in the Grange Life Insurance Company, that thy wife and little ones may have friends when thou art cremated and gathered unto thy fathers.

10. Thou shalt have no Jewish middlemen between thy farm and Liverpool to fatten on thy honest toil, but shalt surely charter thine own ships, and sell thine own produce, and use thine own brains. This is the last and best commandment. On this hang all the law, and profits, and if there be any others they are these.

Choke monopolies, break up rings, vote for honest men, fear God and make money. So shalt thou prosper and sorrow and hard times shall flee away.[4]

4 Oshkosh *Weekly Times*, December 16, 1874.

I. ORIGINAL SOURCES

A. MANUSCRIPTS

Agnew, Samuel Andrew, Diary. Southern Historical Collection, University of North Carolina Library.

Ardrey, William E., Papers. Duke University Library.

Ash, Pauline M., Collection. West Virginia University Library.

Atchison, David Rice, Collection. Western Historical Manuscripts Collection, University of Missouri Library.

Atkeson, Thomas Clark, Papers. West Virginia University Library.

Baird, Sarah G., Papers. Minnesota State Historical Society Library.

Banner Grange (Nebr.) No. 203 Records. Nebraska State Historical Society Library.

Benjamin, John, Papers. Minnesota State Historical Society Library.

Benson-Thompson Family Papers. Duke University Library.

Big Spring Grange (Lawrence, Ind.) No. 1963 Records. Indiana State Library.

Blair, Henry William, Collection. Western Historical Manuscripts Collection, University of Missouri Library.

Blanton, James, Papers. Duke University Library.

Borrors Corners Grange (Jackson Township, Franklin County, Ohio) No. 608, Secretary's Minutes. Ohio State Historical Society Library.

Branch Family Papers. Southern Historical Collection, University of North Carolina Library.

Branchville (Ga.) Grange No. 425, Secretary's Minutes, Branch Family Papers. Southern Historical Collection, University of North Carolina Library.

Bridgers, John L., Jr., Papers. Duke University Library.

Burning Springs Co-operative Association (W. V.) No. 354, Records, Pauline M. Ash Collection. West Virginia University Library.

Center Grange (Yankee Hill, Nebr.) No. 35, Records. Nebraska State Historical Society Library.

Champaign County (Ill.) Pomona Grange, Books. Illinois Historical Survey Collections, University of Illinois Library.

Champaign (Ill.) Grange No. 621, Records. Illinois Historical Survey Collections, University of Illinois Library.

Chase County (Kans.) Manuscripts. Kansas State Historical Society Library.

Church Hill Grange (Ky.) No. 109, Papers. University of Kentucky Library.

Clarke County (Ga.) Pomona Grange No. 101, Records. University of Georgia Library.

Clarke, John P., Papers. West Virginia University Library.

Cottage Grange (Grovewood, S. C.) Papers. Duke University Papers of the Patrons of Husbandry, Duke University Library.

Couchman, George, Family Papers. West Virginia University Library.

Crumpler, W. J., Papers. Duke University Library.

Dickenson, George W., Papers. Duke University Library.

Dimitry, John Bull Smith, Papers. Duke University Library.

Donnelly, Ignatius, Papers. Minnesota State Historical Society Library.

Douglas County (Kans.) Pomona Grange No. 225, Secretary's Minutes. University of Kansas Library.

Dragoon Grange (Osage County, Kans.) No. 331, Secretary's Minutes. Kansas State Historical Society Library.

Excelsior Grange (Palmyra, Nebr.) No. 26, Records. Nebraska State Historical Society Library.

Farmers Co-operative Association (Ill.) Records. Illinois Historical Survey Collections, University of Illinois Library.

Greene County (Ill.) Pomona Grange No. 71, Secretary's Minutes. Illinois State Historical Society Library.

Guilmartin, Lawrence J., and John Flannery, Papers. Duke University Library.

Hanway, James, Scrapbooks. Kansas State Historical Society Library.

Hennepin County (Minn.) Pomona Grange No. 12, Manuscripts. Minnesota State Historical Society Library.

Henry, Jeremiah and Byron V., Papers. Duke University Library.

Hiawatha (Kans.) Grange Secretary's Minutes. Kansas State Historical Society Library.

Hills, J. L. Marginal Comments in copy 2 of Guy B. Horton, *History of the Grange in Vermont*. University of Vermont Library.

Illinois General Assembly Petitions Regarding Railroad Rates, 1869–1874. Illinois State Archives.

Illinois State Grange Records. Illinois State Historical Society Library.

Indiana State Grange Executive Committee Minutes. Indiana State Library.

Indiana State Grange Records. Indiana State Library.

Ives, Lida S., comp. Data Relating to the Patrons of Husbandry. Minnesota State Historical Society Library.

Kansas State Grange Clippings. Kansas State Historical Society Library.

Kelley, Oliver H. House Papers. Minnesota State Historical Society Library.

Keyes, Elisha W., Papers. Wisconsin State Historical Society Library.

Law, William Augustus, Papers. Duke University Library.

Minnehaha Grange (Richfield, Minn.) No. 398, Records. Minnesota State Historical Society Library.

Minnesota Granges Papers. Minnesota State Historical Society Library.

Mississippi Agricultural and Mechanical College Minutes of the Board of Trustees. Mississippi State University Library.

Mississippi Valley Trading Company, Ltd., Papers. Charlotte J. Erikson, ed. Library of the Cooperative Union, Manchester, England, on microfilm at the Minnesota State Historical Society Library.

National Grange. Charter Records of Indiana Granges, 1869–73. Indiana State Library.

National Grange. Applications for Granges in Nebraska with Names of Charter Members, 1872–74. Nebraska State Historical Society Library.

Nebraska State Grange Records. Nebraska State Historical Society Library.

North Star Grange (Minn.) No. 1, Secretary's Minutes. Minnesota State Historical Society Library.

Olive Grange (Harrisville, Ind.) No. 189, Records. Indiana State Historical Society Library.

Osborn, Joseph H., Papers. Oshkosh Public Museum, on microfilm at the Wisconsin State Historical Society Library.

Patrons of Husbandry Manuscripts. University of Virginia Library.

Polk, Leonidas Lafayette, Papers. Southern Historical Collections. University of North Carolina Library.

Pomaria (S. C.) Grange No. 27, Secretary's Minutes. Clemson University Library.

Pratt, Daniel D., Papers. Indiana State Library.

Raleigh Grange (N. C.) No. 17, Papers. Duke University Papers of the Patrons of Husbandry. Duke University Library.

Robson, John N., Papers. Duke University Library.

Rollins, Irwin W., Papers. Minnesota State Historical Society Library.

Samuel, John, Papers. Wisconsin State Historical Society Library.

Saunders, William, Collection. National Grange Office.

Shaw Family Papers. Nebraska State Historical Society Library.

Social Grange (Ill.) No. 1308 Records. Illinois State Historical Society Library.

South Carolina State Grange Papers. Duke University Papers of the Patrons of Husbandry. Duke University Library.

South Carolina State Grange Papers. Clemson University Papers of the Patrons of Husbandry. Clemson University Library.

Taber, Louis J., Papers. Grange Historical Material, Collection of Regional History, Cornell University Library.

University Grange (W. V.) No. 372, Records. West Virginia University Library.

Valley Grange (Gardner, Kans.) No. 312 Secretary's Minutes. University of Kansas Library.

West Butler (Nebr.) Grange No. 476 Papers. Nebraska State Historical Society Library.

Wilmington Grange (Fluvanna County, Va.) Secretary's Minutes. University of Virginia Library.

Winnebago County (Ill.) Pomona Grange Records. Illinois State Historical Society Library.

Wood, Samuel Newitt, Papers. Kansas State Historical Society Library.

B. GOVERNMENT PUBLICATIONS

Aiken, D. Wyatt. "The Grange: Its Origin, Progress, and Educational Purposes." United States Department of Agriculture, *Special Report* 55. Washington, 1883.
Alabama *House Journal*, 1874–82.
Alabama *Senate Journal*, 1874–82.
California *House Journal*, 1873–99.
California *Senate Journal*, 1873–99.
Colorado *House Journal*, 1879–1900.
Colorado *Senate Journal*, 1879–1900.
Connecticut *House Journal*, 1884–99.
Connecticut *Senate Journal*, 1884–99.
Florida *Assembly Journal*, 1874–81.
Florida *Senate Journal*, 1874–81.
Georgia *House Journal*, 1874–82.
Georgia *Senate Journal*, 1874–82.
Hamilton, John. "History of Farmers' Institutes in the United States." United States Department of Agriculture, Office of Experiment Stations, *Bulletin* 174. Washington, 1906.
Holmes, George K. "The Course of Prices of Farm Implements and Machinery for a Series of Years." United States Department of Agriculture, Division of Statistics, *Miscellaneous Series* 18. Washington, 1901.
Illinois Department of Agriculture. *Transactions*, 1872. Springfield, 1873.
Illinois *House Journal*, 1871–99.
Illinois *Laws*, 1869, 1871, 1873.
Illinois Railroad and Warehouse Commission. *Seventh Annual Report*, 1877, Springfield, 1878.
Illinois *Senate Journal*, 1871–99.
Indiana *House Journal*, 1873–99.
Indiana *Senate Journal*, 1873–99.
Iowa *House Journal*, 1873–99.
Iowa *Laws*, 1874.
Iowa *Senate Journal*, 1873–99.
Kansas *House Journal*, 1872–99.
Kansas *Senate Journal*, 1872–99.
Kansas State Board of Agriculture. *Report* XXII. Topeka, 1903.
Kentucky *House Journal*, 1873–85.
Kentucky *Senate Journal*, 1873–85.
Louisiana *House Journal*, 1875–81.
Louisiana *Senate Journal*, 1875–81.
Maine *House Journal*, 1876–99.
Maine *Senate Journal*, 1876–99.
Maryland *House Journal*, 1874–98.
Maryland *Senate Journal*, 1874–98.
Massachusetts *House Journal*, 1874–99.
Massachusetts *Senate Journal*, 1874–99.
Michigan *House Journal*, 1874–99.

Michigan *Senate Journal*, 1874–99.

Michigan State Board of Agriculture. *Annual Report*, 1896–97, 1897–98. Lansing, 1898, 1899.

Minnesota *House Journal*, 1869–99.

Minnesota *Laws*, 1871, 1874.

Minnesota *Senate Journal*, 1869–99.

Mississippi Agricultural and Mechanical College. *Biennial Report*, 1892–93.

Mississippi *House Journal*, 1877–78.

Mississippi *Senate Journal*, 1878.

Missouri *House Journal*, 1873–85.

Missouri *Senate Journal*, 1873–85.

Nebraska *House Journal*, 1875–99.

Nebraska *Senate Journal*, 1875–99.

New Hampshire *Agriculture*, 1893–94. Concord, 1895.

New Hampshire *House Journal*, 1875–99.

New Hampshire *Senate Journal*, 1875–99.

New Jersey *Minutes of the Assembly*, 1875–99.

New Jersey *Senate Journal*, 1875–99.

New York *Assembly Journal*, 1875–99.

New York *Senate Journal*, 1875–99.

North Carolina *House Journal*, 1874–81.

North Carolina *Senate Journal*, 1874–81.

Ohio *House Journal*, 1873–99.

Ohio *Senate Journal*, 1873–99.

Oregon *House Journal*, 1874–99.

Oregon *Senate Journal*, 1874–99.

Pennsylvania *House Journal*, 1874–99.

Pennsylvania *Senate Journal*, 1874–99.

South Carolina *House Journal*, 1874–82.

South Carolina *Senate Journal*, 1874–82.

Tennessee *House Journal*, 1875–81.

Tennessee *Senate Journal*, 1875–81.

Texas *House Journal*, 1875–89.

Texas *Senate Journal*, 1875–89.

True, Alfred C. "A History of Agricultural Education in the United States, 1785–1925." United States Department of Agriculture, *Miscellaneous Publication* 36. Washington, 1929.

————. "A History of Agricultural Experimentation and Research in the United States, 1607–1925." United States Department of Agriculture, *Miscellaneous Publication* 251. Washington, 1937.

————. "A History of Agricultural Extension Work in the United States, 1785–1923." United States Department of Agriculture, *Miscellaneous Publication* 15. Washington, 1928.

United States Census Office. *Twelfth Census* 1900, *Agriculture*, Part I. Washington, 1902.

United States Commissioner of Agriculture. *Report*, 1867. Washington, 1868.

United States *Congressional Record*, 48th–52nd Congs.

United States Department of Agriculture. "Cotton and Cottonseed: Acreage,

Yield, Production, Disposition, Price, and Value, by States, 1866–1952."
USDA, Agricultural Marketing Service, *Statistical Bulletin* 164. Washington, 1955.

————. "Fluctuations in Crops and Weather, 1866–1948." USDA, *Statistical Bulletin* 101. Washington, 1951.

————. "Potatoes: Acreage, Production, Value, Farm Disposition, Jan. 1 Stocks, 1866–1950." USDA, Bureau of Agricultural Economics, *Statistical Bulletin* 122. Washington, 1953.

————. *Yearbook*, 1899. Washington, 1900.

United States Senate. "Report of the Select Committee on Transportation—Routes to the Seaboard." *Senate Report* 307, Washington, 1874.

Vermont *House Journal*, 1876–98.

Vermont *Senate Journal*, 1876–98.

Virginia *House Journal*, 1874–83.

Virginia *Senate Journal*, 1874–83.

Washington *House Journal*, 1889–99.

Washington *Senate Journal*, 1889–99.

West Virginia *House of Delegates Journal*, 1875–1900.

West Virginia *Senate Journal*, 1875–1900.

Wisconsin *Assembly Journal*, 1872–99.

Wisconsin *Laws*, 1874.

Wisconsin *Senate Journal*, 1872–99.

C.　NEWSPAPERS AND MAGAZINES

Albany *Cultivator & Country Gentleman*, 1875–97.

Alexandria (Va.)*Granger*, 1876.

Amherst (N. H.) *Farmers' Cabinet*, 1878–81.

Athens (Ga.) *Southern Cultivator*, 1872–78.

Atlanta *Georgia Grange*, 1877.

Atlanta *Wilson's Herald of Health, and Farm and Household Help*, 1873–74. *Wilson's Herald of Health and Atlanta Business Review and The Rural Southerner* and *Wilson's Herald of Health*. Title fluctuated.

Augusta *Southern Cultivator and Dixie Farmer*, 1886–88.

Baltimore *American Farmer*, 1872–82.

Bangor, (Me.) *Dirigo Rural*, 1874–82.

Bloomington (Ill.) *Anti-Monopolist*, 1873.

Bloomington (Ill.) *Appeal*, 1876.

Booneville (Ark.) *Enterprise*, 1875–77.

Brownville (Nebr.) *Advertiser*, 1872–76.

Brownville (Nebr.) *Nemaha County Granger*, 1876–80.

Charleston *Rural Carolinian*, 1871–76.

Charleston *West Virginia Farm Reporter*, 1897.

Charleston *West Virginia Farm Review*, 1899.

Charlotte *Southern Home*, 1871–76.

Chicago *Christian Cynosure*, 1874–76.

Chicago *Industrial Age*, 1873–77.

Chicago *Prairie Farmer*, 1872–81.

Chicago *Tribune*, 1870–74.
Chicago *Western Rural*, 1875–84.
Cincinnati *American Grange Bulletin and Scientific Farmer*, 1896–99.
Clarksville (Tenn.) *Tobacco Leaf*, 1873–74.
Cleveland *Ohio Farmer*, 1870–1900.
Columbus (Miss.) *Patron of Husbandry*, 1879–83.
Concord *New Hampshire Agriculturist and Patron's Journal*, 1895–96.
Denver *Colorado Farmer*, 1884–85.
Denver *Rocky Mountain Weekly News*, 1874.
Des Moines *Iowa Homestead and Western Farm Journal*, 1870–75.
Elmira (N. Y.) *Husbandman*, 1891.
Fayetteville (Ark.) *Democrat*, 1873–76.
Fort Smith (Ark.) *Herald*, 1872–76.
Fort Smith (Ark.) *Western Independent*, 1872–77.
Friendship (Wis.) *Adams County Press*, 1873–76.
Galena (Ill.) *Industrial Press*, 1876–77.
Garnett (Kans.) *Weekly Journal*, 1874.
Greenville (S. C.) *Cotton Plant*, 1887–88.
Hillsboro (Ill.) *Anti-Monopolist*, 1875.
Humboldt (Tenn.) *Grange Journal*, 1876.
Indianapolis *Indiana Farmer*, 1874–88.
Indianapolis (and Muncie) *Hoosier Patron*, 1876.
Lawrence *Spirit of Kansas*, 1873–77.
Lincoln *Nebraska Farmer*, 1877–99.
Lincoln *Nebraska Patron*, 1874–76. (Title changed to *Independent Patron*
     in 1876.)
Little Rock *Arkansas Grange*, 1874.
Louisville (and Lexington) *Farmers Home Journal*, 1870–89.
Macomb *Illinois Granger*, 1875.
Macon (Ga.) (and Memphis) *Southern Farm and Home*, 1870–74.
Macon *Missouri Granger*, 1874–75.
Madison *Western Farmer*, 1870–74.
Milwaukee *Sentinel*, 1873.
Minneapolis *Evening Times and News*, 1873.
Minneapolis *Farmers' Union*, 1867–74.
Mobile *Rural Alabamian*, 1872–73.
Nashville *Rural Sun*, 1872–73.
New York *American Agriculturist*, 1870–1900.
New York *Rural New Yorker*, 1871–1900.
New York *Times*, 1873–80.
Omaha *Central Union Agriculturist and Missouri Valley Farmer*, 1869–73.
Omaha *Rural Nebraska*, 1879–85.
Oregon (Ill.) *Ogle County Grange*, 1875.
Oshkosh (Wis.) *Weekly Times*, 1873–77.
Portsmouth *Virginia Granger*, 1880–84.
Raleigh *North Carolina Farmer*, 1877–82.
Raleigh (N. C.) *State Agricultural Journal*, 1873–75.
Richmond *Virginia Patron*, 1877.

St. Louis *Colman's Rural World*, 1870–93.
St. Louis *Journal of Agriculture*, 1896–1900.
St. Paul *Anti-Monopolist*, 1874–75.
St. Paul (and Red Wing) *Grange Advance*, 1874–77.
St. Paul *Minnesota Monthly*, 1869.
San Francisco *Pacific Rural Press*, 1871–1900.
Spartanburg *Carolina Spartan*, 1887–88.
Springfield, Ohio, *Grange Visitor and Farmers' Monthly Magazine*, 1875–76.
Starkville (Miss.) *Southern Live-Stock Journal*, 1881–90.
Taylorville (Ill.) *Christian County Farmers' Journal*, 1878, 1880.
Topeka *Capital*, 1881.
Topeka *Commonwealth*, 1883–88. (Title subsequently changed to *Capital-Commonwealth*.)
Topeka *Kansas Daily State Journal*, 1885.
Topeka (and Leavenworth) *Kansas Farmer*, 1870–92.
Waukon, Iowa, *Standard*, 1873–74.
Wilmington *Carolina Farmer*, 1872, 1877–79.

D.   GRANGE AND MISCELLANEOUS ORGANIZATION PROCEEDINGS

Alabama State Grange, 1874–75, 1888.
Arkansas State Grange, 1877.
Association of American Agricultural Colleges and Experiment Stations, 1889–90.
Association of Land-Grant Colleges and Universities, 1926.
Atlantic Council, Patrons of Husbandry, 1873.
California State Grange, 1874, 1885–88, 1890–94, 1897–99.
Colorado State Grange, 1888–89, 1892.
Connecticut State Grange, 1885–97.
Delaware State Grange, 1888–92, 1897, 1899.
Florida State Grange, 1875.
Georgia State Grange, 1873, 1875, 1886.
Illinois State Grange, 1875–76, 1889–99.
Indiana State Grange, 1874, 1877–99.
Iowa State Grange, 1872–77, 1883–99.
Kansas State Grange, 1875–1900.
Kentucky State Grange, 1887.
Louisiana State Grange, 1875.
Maine State Grange, 1874–75, 1897–99.
Maryland State Grange, 1874–91, 1894–98.
Massachusetts State Grange, 1882–99.
Michigan State Grange, 1875–80, 1891.
Minnesota State Grange, 1890–99.
Mississippi State Grange, 1875.
Missouri State Grange, 1875–76, 1886–92.
Montana Territorial Grange, 1875.
National-American Woman Suffrage Association, 1884–99.

National Grange, 1873–99.
Nebraska State Grange, 1874, 1887, 1891.
New Hampshire State Grange, 1874–84, 1886–95, 1898–99.
New Jersey State Grange, 1877–79, 1891.
New York State Grange, 1875–76, 1889–97.
North Carolina State Grange, 1873–78, 1882–84. 1886–87.
Northern Virginia District Grange, 1885.
Ohio State Grange, 1874–80, 1890–91, 1893–94, 1896–99.
Oregon State Grange, 1873, 1875–78, 1881–83, 1885–88, 1891.
Pennsylvania State Grange, 1886–87.
Rhode Island State Grange, 1894–95.
Society for the Promotion of Agricultural Sciences, 1907.
South Carolina State Grange, 1874, February 1875–December 1875, 1877, February 1880–December 1880, 1881–84, 1886–89.
Tennessee State Grange, 1875.
Texas Co-operative Association, Patrons of Husbandry, 1880–97.
Texas State Grange, 1874, 1877–82, 1884–86, 1889, 1892–95, 1897, 1899.
Vermont State Grange, 1872–75, 1886, 1891, 1893, 1897.
Virginia State Grange, 1874–76, 1878–79, 1886, 1890.
Washington State Grange, 1889, 1899.
West Virginia State Grange, 1891, 1893, 1895, 1897, 1899.
Wisconsin State Agricultural Society, 1883.
Wisconsin State Grange, 1874–99.

E. PRIMARY BOOKS

Atkeson, Thomas C. *History of the Declaration of Purposes of the Grange.* n.d., n.p.
———. *Outlines of Grange History.* Washington, 1928.
Atkeson, Thomas C., and Mary Meek Atkeson. *Pioneering in Agriculture: One Hundred Years of American Farming and Farm Leadership.* New York, 1937.
Butterfield, Kenyon. *Chapters in Rural Progress.* Chicago, 1908.
Buell, Jennie. *One Woman's Work for Farm Women: The Story of Mary A. Mayo's Part in Rural Social Movements.* Boston, 1908.
———. *The Grange Master and the Grange Lecturer.* New York, 1921.
Carr, Ezra S. *The Patrons of Husbandry of the Pacific Coast.* San Francisco, 1875.
Connecticut State Grange, *The Connecticut Granges: An Historical Account of the Rise and Growth of the Patrons of Husbandry.* New Haven, 1900.
Darrow, J. Wallace. *Origin and Early History of the Order of Patrons of Husbandry in the United States.* Chatham, N.Y., 1904.
Dominion Grange. *History of the Grange in Canada.* Toronto, 1876.
Garland, Hamlin. *A Son of the Middle Border.* New York, 1917.
Grosh, Aaron B. *Mentor in the Granges and Homes of Patrons of Husbandry.* New York, 1876.
Gustin, M. E. *An Expose of the Grangers.* Dayton, 1875.

Kelley, Oliver H. *Origin and Progress of the Order of the Patrons of Husbandry in the United States: A History from 1866 to 1873.* Philadelphia, 1873.

Martin, Edward Winslow (James D. McCabe). *History of the Grange Movement.* Chicago, 1874.

National Christian Association. *A Brief History of the National Christian Association.* Chicago, 1874.

Smedley, A. B. *The Principles and Aims of the Patrons of Husbandry.* Burlington, Iowa, 1874.

Smith, Stephen R. *Grains for the Grangers: Discussing All Points Bearing upon the Farmers' Movement for the Emancipation of White Slaves from the Slave-Power of Monopoly.* Chicago, 1873.

Wells, John G. *The Grange: A Study in the Science of Society Practically Illustrated by Events in Current History.* New York, 1874.

World Almanac, 1893, 1897. New York, 1893, 1897.

F. PRIMARY ARTICLES

Adams, Charles F., Jr. "The Granger Movement." *North American Review,* CXX (April, 1875), 394–424.

Bobbitt, T. N. "My Recollection of the Early Grange in Nebraska." *Nebraska History and Record of Pioneer Days,* V (January, 1923), 13–14.

Buell, Jennie. "The Educational Value of the Grange." *Business America,* XIII (January, 1913), 50–54.

Butterfield, Kenyon L. "A Significant Factor in Agricultural Education." *Educational Review,* XX (1901), 301–306.

————. "The Grange." *The Forum,* XXXI (April, 1901), 231–42.

Coulter, John Lee. "Organization Among the Farmers of the United States." *Yale Review,* XVIII (November, 1909), 273–98.

Dodge, John R. "The Discontent of the Farmer." *Century,* XXI (January, 1892), 447–56.

Drew, Frank M. "The Present Farmer's Movement." *Political Science Quarterly,* VI (June, 1891), 282–310.

Edsall, James K. "The Granger Cases and the Police Power." American Bar Association, *Reports,* X (1887), 288–316.

Emerick, C. F. "An Analysis of the Agricultural Discontent in the United States." *Political Science Quarterly,* XI & XII (September, December, 1896, March, 1897), 433–63, 601–39, 93–127.

Flagg, Willard C. "The Farmers' Movement in the Western States." *Journal of Social Science,* VI (July, 1874), 100–15.

Foster, Florence J. "The Grange and the Co-operative Enterprises in New England." American Academy of Political and Social Science, *Annals,* IV (March, 1894), 798–805.

"The Granger Collapse." *The Nation,* XXII (January 27, 1876), 57–58.

"The Granger Method of Reform," *The Nation,* XIX (July 16, 1874), 36–37.

Hadley, Arthur T. "Yale University." *Harper's New Monthly Magazine,* LXXXVIII (April, 1894), 764–72.

Howland, Marie. "The Patrons of Husbandry." *Lippincott's Magazine*, XII (September, 1873), 338–42.

Kelley, Oliver H. "Early Struggles of the Grange." *American Grange Bulletin*, XXXIV (September 1, 1904), 8–9.

————. "Grange History; Personal Reminiscences." *American Grange Bulletin*, XXXV (August 31, 1905), 9.

Pierson, Charles W. "The Outcome of the Granger Movement." *Popular Science Monthly*, XXXII (January, 1888), 368–73.

————. "The Rise of the Granger Movement." *Popular Science Monthly*, XXXII (December, 1887), 199–208.

Swalm, Pauline. "The Granges of the Patrons of Husbandry." *Old and New*, VIII (1873), 96–100.

Titus, Frank. "The Grange Problem." *Kansas Magazine*, IV (September, 1873), 282–85.

Whitehead, Mortimer. "The Grange in Politics." *American Journal of Politics*, I (August, 1892), 113–23.

G. MISCELLANEOUS

American Cooperative Union. *Manuel of Practical Cooperation by Thomas D. Worrall, Louisville, 1875*. Louisville, 1875.

California State Grange. *Organization of the California State Grange at Napa City, July 15th, 1873: Declaration of Purposes, Plan of Operations, Preliminary Procedures, Etc.* San Francisco, 1873.

Cedar Mountain (Va.) Grange No. 353. *Constitution.* Culpeper, 1875.

Cramer, J. A. *The Patron's Pocket Companion in Four Parts.* Cincinnati, 1875.

Direct Trade Union. Circular by A. H. Colquitt. *To the Patrons of Husbandry of the Cotton States.* Atlanta, 1875.

Donnelly, Ignatius. *Facts for the Granges.* n.p., 1873.

Dunn County (Wis.) Pomona Grange. *Constitution and By-Laws of the Cooperative Council of Dunn County Patrons of Husbandry.* Neenah, Wis., 1875.

Elm Creek (Tex.) Cooperative Association. *Charter and By-Laws.* Sequin, Tex., 1891.

Fond Du Lac County (Wis.) Pomona Grange No. 9. *Constitution and By-Laws.* Fond DuLac, 1877.

Greene County (Ill.) Pomona Grange No. 71. *By Laws.* Greenfield, Ill., 1891.

Hall, Z. M., Wholesale Grocer. Patrons of Husbandry, *Catalogue.* January 20, 1875, October 5, 1875, Chicago, 1875.

Hamilton, Robert A. *Mission of the Patrons of Husbandry; Birds and Insects: Two Addresses before Young's Cross Roads Grange, Granville County, North Carolina.* Petersburg, Va., 1875.

Hudson, Joseph K. *The Patrons' Handbook; for the Use and Benefit of the Order of the Patrons of Husbandry.* Topeka, 1874.

Illinois State Grange. *By-Laws Adopted December 16, 1875.* Sterling, Ill., 1876.

Indiana State Grange Purchasing Agency. *Circular 7.* Indianapolis, n.d.

Jevne, C., Wholesale Grocer. *Wholesale Prices Catalogue*. October 16, 1876, January 22, 1877, Chicago, 1876, 1877.

Kansas State Grange. *Constitution and By-Laws*. Lawrence, n.d.

————. *Report of the Standing Committee on Education*. 1884, n. p., n. d.

————. *Report of the Standing Committee on Education* (December 14, 1881). F. G. Adams, "Education for Farmers' Children." Topeka, 1882.

Kenoma Grange, Anderson County, Kansas. Circular entitled, *Be Warned*, December 22, 1873.

Kentucky State Grange. *Constitution and By-Laws*. Lexington, 1875.

Maine State Grange. *Constitution and By-Laws*. Standish, Maine, 1896.

Maryland State Grange. *By-Laws and Regulations for Formation of District or County Granges*. Baltimore, 1875.

Massachusetts State Grange. *Constitution and By-Laws*. Hudson, Mass., 1896.

————. William H. Earle's Address on *Cooperation*, delivered December 22, 1880.

————. *Quarterly Bulletin*. n.p., January, 1876.

Midway Grange (Va.) No. 217, *By-Laws*. Richmond, 1874.

Minnesota State Grange. *Constitution Adopted February, 1873*. Lake City, Minn., 1873.

Mississippi State Grange. *By-Laws*. n.p., n.d.

————. *The State Grange and A. & M. College*. n.p., n.d.

Mississippi Valley Trading Company, Ltd. *International Co-operation*. n.p., 1875.

————. *Memorandum and Articles of Association*. Manchester, England, 1875.

Missouri State Grange. *Constitution and By-Laws*. Knob Noster, Mo., 1874.

National Grange. *Bryan Fund Publications, 2–4, 6, 8, 10*. Washington, 1872.

————. *Grasshopper Circular*. Plattsmouth, Nebr., 1874.

————. *Manual of Subordinate Granges*. Philadelphia, 1874.

————. *Rules for Patrons' Co-Operative Associations of the Order of Patrons of Husbandry, and Directions for Organizing Such Associations Recommended by the National Grange, November, 1875*. Louisville, 1876.

————. *Songs of the Grange*. Philadelphia, 1874.

Nebraska State Grange. *Constitution*. Lincoln, 1874.

New Hampshire Grange Fair Association. *Official Premium List and Rules and Regulations of Sixth Annual Exhibition*. n.p., n.d.

New Hampshire State Grange. *By-Laws Adopted December, 1876*. Claremont, 1877.

Nodaway County (Mo.) Pomona Grange. *Constitution*. Maryville, Mo., 1873.

————. *Constitution and By-Laws*. Maryville, Mo., 1885.

Northern Virginia District Grange. Broadside entitled *Subscription Supper to State Grange, February 17th, 1882 at Green's Mansion House, Alexandria, Va.*

Ohio State Grange. *Hints and Helps to Profit and Pleasure in the Grange with Topics for Discussion and Programmes for Meetings*. Springfield, Ohio, 1881.

————. Mutual Protection Association of the Patrons of Husbandry, *By-Laws*. Kenton, Ohio, 1876.

_____. *Confidential Trade Circular*, 1895, 1898, 1899. Springfield, 1895, 1898, 1899.

Paine, E. T. *The Direct Trade Union: Its Objects and Advantages*. Atlanta, 1874.

Patrons Aid Association, Orangeburg, S. C. *By-Laws*. Charleston, 1877.

Patrons Aid Society, Elmira, N. Y. *By-Laws*. Elmira, 1876.

_____. *Sixth Annual Report*. Elmira, 1881.

Patrons of Husbandry. *Prospectus of the First Annual Inter-State Farmers' Summer Encampment*. Spartanburg, 1887.

Rock County (Wisc.) Pomona Grange No. 5. *Constitution and By-Laws*. Janesville, 1877.

South Carolina State Grange. *By-Laws*. Charleston, 1873.

Texas State Grange. *Annual Address of Worthy Master A. J. Rose of the Texas State Grange, Delivered at Marlin, Texas, August 9, 1887*. Dallas, 1887.

_____. *Annual Exhibit of the Texas State Grange Fair*, 1890, 1891. Temple, 1890, 1891.

_____. *Constitution*. Waco, 1874.

_____. *Constitution and Declaration of Purposes of the National Grange, Patrons of Husbandry, together with the Constitution of the Texas State Grange*. Galveston, 1885.

Tipton, Iowa. Cooperative Association. *Articles of Incorporation and Constitution*. Tipton, 1874.

Virginia State Grange, Business Bureau. *Confidential Circular*, July, 1877. Lynchburg, 1877.

Virginia State Grange. *Communication . . . Expressing their Opinions and Wishes in Reference to the Inspection of Tobacco*. Richmond, 1876.

_____. Executive Committee. *Confidential Circular*, 6. n.p., 1875.

Wabash Grange (Ind.) No. 1596. *By-Laws*. Covington, 1874.

Watertown (Conn.) Grange No. 122. *By-Laws*. n.p., 1892.

Wisconsin State Grange. *Bulletin*, 1875–84. Madison, 1875–84.

_____. *Constitution and By-Laws*. Oshkosh, 1874.

_____. *Grange Agency Catalogue of 1877*. Milwaukee, 1877.

Worcester Grange. *Constitution and By-Laws*. Worcester, Mass., 1883.

Worrall, Thomas D. *Direct Trade Between Great Britain and the Mississippi Valley*. Manchester, England, 1875.

Wright, J. W. A. *Addresses on Rochdale and Grange Co-operation and Other Grange Topics*. San Francisco, 1877.

II. SECONDARY MATERIALS

A. BOOKS

Adams, Herbert B., ed. *History of Cooperation in the United States*. Baltimore, 1888.

Allen, Gay W. *William James; A Biography*. New York, 1967.

Allen, Leonard L. *History of New York State Grange*. Watertown, 1934.

Archambault, Reginald D. *Dewey on Education: Appraisals.* New York, 1966.

Baker, Melvin C., *Foundations of John Dewey's Educational Theory.* New York, 1955.

Barnes, C. R. "The Department of Agriculture," in *Forty Years of the University of Minnesota,* edited by E. Bird Johnson. Minneapolis, 1910.

Beal, W. J. *History of the Michigan Agricultural College and Biographical Sketches of Trustees and Professors.* East Lansing, 1915.

Benedict, Murray R. *Farm Policies of the United States, 1790–1950.* New York, 1953.

Benson, Lee. *Merchants, Farmers, and Railroads; Railroad Regulation and New York Politics, 1850–1887.* Cambridge, 1955.

Bettersworth, John K. *People's College: A History of Mississippi State.* University, Alabama, 1953.

Bidwell, Percy W., and John L. Falconer. *History of Agriculture in the Northern United States, 1620–1860.* New York, 1941.

Boss, Andrew. *The Early History and Background of the School of Agriculture at University Farm, St. Paul.* n.p., 1941.

Brenckman, Frederick C. *History of the Pennsylvania State Grange.* Harrisburg, 1949.

Brooks, Robert P. *The Agrarian Revolution in Georgia, 1865–1912.* Madison, 1914.

Brunner, Edmund deS., and E. Hsin Pao Yang. *Rural America and the Extension Service: A History and Critique of the Cooperative Agricultural and Home Economics Extension Service.* New York, 1949.

Buck, Paul H. *The Road to Reunion: 1865–1900.* New York, 1937.

Buck, Solon J. *The Granger Movement; A Study of Agricultural Organization and Its Political, Economic and Social Manifestation, 1870–1880.* Cambridge, Mass., 1933.

Chittenden, Russell H. *History of the Sheffield Scientific School of Yale University, 1846–1922.* 2 vols.; New Haven, 1928.

Colman, Gould P. *Education & Agriculture—A History of the New York State College of Agriculture at Cornell University.* Ithaca, 1963.

Commons, John R., and others, eds. *A Documentary History of American Industrial Society.* 10 vols.; Cleveland, 1910.

Crawford, Harriet A. *The Washington State Grange, 1889–1924.* Portland, 1940.

Cremin, Lawrence A. *The Transformation of the School: Progressivism in American Education, 1876–1957.* New York, 1961.

Curti, Merle, and Vernon Carstensen. *The University of Wisconsin.* 2 vols.; Madison, 1949.

Day, Clarence A. *Farming in Maine, 1860–1940.* Orono, Maine, 1963.

Dunaway, Wayland F. *History of Pennsylvania State College.* State College, 1946.

Fite, Gilbert C. *The Farmers' Frontier, 1865–1900.* New York, 1966.

Fleming, Walter L. *Louisiana State University, 1860–1896.* Baton Rouge, 1936.

Fletcher, Stevenson W. *Pennsylvania Agriculture and Country Life, 1840–1940.* Harrisburg, 1955.

Flexner, Eleanor. *Century of Struggle: The Woman's Rights Movement in the United States.* Cambridge, Mass., 1959.

Fossum, Paul R. *The Agrarian Movement in North Dakota.* Baltimore, 1925.

Franklin, John H. *Reconstruction After the Civil War.* Chicago, 1961.

Fuller, Wayne E. *RFD—The Changing Face of Rural America.* Bloomington, Ind., 1964.

Gardner, Charles M. *The Grange—Friend of the Farmer.* Washington, 1949.

Garraty, John A. *The New Commonwealth, 1877–1890.* New York, 1968.

Gates, Paul W. *Agriculture and the Civil War.* New York, 1965.

Geiger, George R. *John Dewey in Perspective.* New York, 1958.

Glover, W. H. *Farm and College; the College of Agriculture of the University of Wiconsin.* Madison, 1952.

Gray, James. *The University of Minnesota.* Minneapolis, 1951.

Haynes, Frederick E. *James Baird Weaver.* Iowa City, 1919.

Hicks, John D. *The Populist Revolt.* Minneapolis, 1931.

Hill, William F. *A Brief History of the Grange Movement in Pennsylvania.* Chambersburg, Pennsylvania, 1923.

Hofstadter, Richard. *The Age of Reform, from Bryan to F. D. R.* New York, 1955.

Holden, Ellsworth A., ed. *Souvenir: National Grange in Michigan.* Lansing, 1902.

Hopkins, Janet W. *History of Rhode Island State Grange.* n.p., 1939.

Horton, Guy B. *History of the Grange in Vermont.* Montpelier, 1926.

Hunt, Robert L. *A History of Farmer Movements in the Southwest, 1873–1925.* College Station, Tex., 1935.

Jarchow, Merrill E. *The Earth Brought Forth: A History of Minnesota Agriculture to 1885.* St. Paul, 1949.

Kansas State Grange. Report of Historical Committee, *The Grange Movement in Kansas.* n.p., 1952.

Kinnear, Duncan L. *The First 100 Years: A History of Virginia Polytechnic Institute and State University.* Blacksburg, 1972.

Kinnison, William A. *Building Sullivant's Pyramid: An Administrative History of the Ohio State University, 1870–1907.* Columbus, 1970.

Kirwan, Albert D. *Revolt of the Rednecks: Mississippi Politics, 1876–1925.* Lexington, 1951.

Knoblauch, H. C., et al. *State Agricultural Experiment Stations: A History of Research Policy and Procedure,* USDA Misc. Pub. 904, Washington, 1962.

Kolko, Gabriel. *Railroads and Regulation, 1877–1916.* Princeton, 1965.

Kraditor, Aileen S. *The Ideas of the Woman Suffrage Movement, 1890–1920.* New York, 1965.

Kuhn, Madison. *Michigan State: The First Hundred Years, 1855–1955.* East Lansing, 1955.

Latta, W. C. *Outline History of Indiana Agriculture.* Lafayette, 1938.

Lemmer, George F. *Norman J. Colman and "Colman's Rural World"—A Study in Agricultural Leadership.* Columbia, Mo., 1953.

Lindstrom, David E. *American Farmers' and Rural Organizations.* Champaign, Ill., 1948.

Miller, George H. *Railroads and the Granger Laws.* Madison, 1971.

Minnesota State Grange. *History, State Grange of Minnesota.* n.p., 1947.

Moore, Edward C. *American Pragmatism: Pierce, James and Dewey.* New York, 1961.

Moores, Richard G. *Fields of Rich Toil: the Development of the University of Illinois College of Agriculture.* Urbana, 1970.

National Grange. *Legal and Economic Influence of the Grange, 1867–1967.* Washington, 1967.

Noblin, Stuart. *Leonidas Lafayette Polk—Agrarian Crusader.* Chapel Hill, 1949.

Nordin, Dennis S. *A Preliminary List of References for the History of the Granger Movement.* Davis, Calif., 1967.

Nugent, Walter T. K. *The Tolerant Populists: Kansas, Populism and Nativism.* Chicago, 1963.

Ohio State Grange. *Diamond Jubilee History, Ohio State Grange. 1872–1947.* Columbus, 1947.

_____. *Golden Jubilee History, 1872–1922.* n.p., n.d.

Onondaga Pomona Grange. *Historical Directory, 1883–1915.* Oswego, N.Y., 1915.

O'Rourke, Alice A. "Cooperative Marketing in McLean County," *Journal of the Illinois State Historical Society,* LXIV (Summer, 1971).

Paine, Arthur E. *The Granger Movement in Illinois.* Urbana, 1904.

Peterson, James A. *Grange of Illinois.* Chicago, 1956.

Pollack, Norman. *The Populist Response to Industrial America: Midwestern Populist Thought.* Cambridge, Mass., 1962.

Potter, Beatrice. *The Cooperative Movement in Great Britain.* London, 1930.

Ridge, Martin. *Ignatius Donnelly: The Portrait of a Politician.* Chicago, 1962.

Robinson, W. L. *The Grange, 1867–1967: First Century of Service and Evolution.* Washington, 1966.

Ross, Earle D. *Democracy's College; the Land-Grant Movement in the Formative Stage.* Ames, Iowa, 1942.

Saby, Rasmus S. "Railroad Legislation in Minnesota, 1849–1875," *Minnesota Historical Collections,* XV (May, 1915), 1–188.

Saloutos, Theodore. *Farmer Movements in the South, 1865–1933.* Berkeley, 1960.

Scheele, Carl H. *A Short History of the Mail Service.* Washington, 1970.

Schmidt, Louis B. "Farmers' Organizations," *A Century of Farming in Iowa, 1846–1946.* Ames, Iowa, 1946.

Schonberger, Howard B. *Transportation to the Seaboard: the "Communication Revolution" and American Foreign Policy 1860–1900.* Westport, 1971.

Scott, Roy V. *The Agrarian Movement in Illinois, 1880–1896.* Urbana, Ill., 1962.

Scott, Roy V. *The Reluctant Farmer: The Rise of Agricultural Extension to 1914.* Urbana, 1970.

Shannon, Fred A. *The Farmer's Last Frontier.* New York, 1945.

Smith, Ralph W. *The History of the Iowa State Grange, Patrons of Husbandry, from 1868 to 1946.* Manchester, Iowa, 1946.

Snyder, Ralph. *We Kansas Farmers: Development of Farm Organizations and Cooperative Associations in Kansas.* Topeka, 1953.

Stemmons, Walter. *Connecticut Agricultural College—A History.* Storrs, Conn., 1931.

Taylor, Carl C. *The Farmers' Movement, 1620–1920.* New York, 1953.

Trump, Fred. *The Grange in Michigan.* Grand Rapids, 1963.

Tyler, Alice Felt. *The Foreign Policy of James G. Blaine.* Minneapolis, 1927.

Tyler, Helen E. *Where Prayer and Purpose Meet: The WCTU—1874–1949.* Evanston, 1949.

U.S. *Statutes at Large.* Washington.

Wiest, Edward. *Agricultural Organization in the United States.* Lexington, 1923.

Wood, Louis A. *A History of Farmers' Movements in Canada.* Toronto, 1924.

Woodward, Carl R. *The Development of Agriculture in New Jersey, 1640–1880.* New Brunswick, 1927.

Working, Daniel. *Colorado State Grange, 1874–1924.* Denver, 1924.

_____, and Alvin T. Steinel. *History of Agriculture in Colorado, 1858 to 1926.* Fort Collins, 1926.

Yearley, Clifton K., Jr. *Britons in American Labor: A History of the Influence of the United Kingdom Immigrants on American Labor, 1820–1914.* Baltimore, 1957.

B. ARTICLES

Aldrich, Charles. "The Repeal of the Granger Law in Iowa." *Iowa Journal of History and Politics,* III (April, 1905), 256–70.

Ander, O. Fritiof. "The Immigrant Church and the Patrons of Husbandry." *Agricultural History,* VIII (October, 1934), 155–68.

Anderson, W. A. "The Granger Movement in the Middle West with Special Reference to Iowa." *Iowa Journal of History and Politics,* XXII (January, 1924), 3–51.

Backstrom, Philip N. "The Mississippi Valley Trading Company: A Venture in International Cooperation 1875–1877." *Agricultural History,* XLVI (July, 1972), 425–37.

Ball, Joseph. ["Beginnings of the Granger Movement in Minnesota,"] Braham *Minnesota Grange Gleaner,* May 20, June 17, July 15, 1940.

Barns, William D. "Farmers vs. Scientists: The Grange, the Farmers' Alliance and the West Virginia Agricultural Experiment Station." West Virginia Academy of Science, *Proceedings,* XXXVII (1965), 197–206.

_____. "Oliver Hudson Kelley and the Genesis of the Grange: A Reappraisal." *Agricultural History,* XLI (July, 1967), 229–42.

_____. "Record of Achievement; Historical Sketch of the Grange in West Virginia." *Farmer of West Virginia,* II (November, 1955), 11, 16, 21.

_____. "The Influence of the West Virginia Grange upon Public Agricultural Education of College Grade, 1873–1914." *West Virginia History,* IX (January, 1948), 128–57.

_____. "The Influence of the West Virginia Grange upon Public Agricultural Education of Less than College Grade, 1873–1914." *West Virginia History*, X (October, 1948), 5–24.

Beddie, Ruth D. "The North Star Grange." *Minnesota History*, XIX (March, 1938), 98–99.

Beinhauer, Myrtle. "Development of the Grange in Iowa." *Annals of Iowa*, XXXIV (April, 1959), 597–618.

Bennett, James D. "Some Notes on Christian County, Kentucky, Grange Activities." Kentucky Historical Society, *Register*, LXIV (July, 1966), 226–34,

Betts, John R. "Agricultural Fairs and the Rise of Harness Racing." *Agricultural History*, XXVII (April, 1953), 71–75.

Blegen, Theodore C. "The Farmer's Crusade in Minnesota." *Minnesota Alumni Weekly*, XXXII (May 20, 1933), 511–12.

"Brief History of the Grange." *Farmer's Elevator*, XXXIII (June, 1938).

Briggs, John E. "The Grasshopper Plagues in Iowa." *Iowa Journal of History and Politics*, XIII (July, 1915), 349–91.

Buck, Solon J. "Agricultural Organizations in Illinois, 1870–1880." Illinois State Historical Society, *Journal*, III (April, 1910), 10–23.

Campbell, Ballard. "The Good Roads Movement in Wisconsin, 1890–1911." *Wisconsin Magazine of History*, XLIX (Summer, 1966), 273–93.

Campbell, C. A. "Grange, Its Work and Ideals." *New England Magazine*, (April, 1910), 184–91.

Carstensen, Vernon. "The Genesis of an Agricultural Experiment Station." *Agricultural History*, XXXIV (January, 1960), 13–20.

Cerny, George. "Cooperation in the Midwest in the Granger Era, 1869–1875." *Agricultural History*, XXXVII (October, 1963), 187–205.

Crawford, Harriet A. "Grange Attitudes in Washington, 1889–1896." *Pacific Northwest Quarterly*, XXX(July, 1939), 243–74.

Daland, Robert T. "Enactment of the Potter Law." *Wisconsin Magazine of History*, XXXIII (September, 1949), 45–54.

Davis, Granville D. "The Granger Movement in Arkansas." *Arkansas Historical Quarterly*, IV (Winter, 1945), 340–52.

Detrick, C. R. "The Effects of the Granger Acts." *The Journal of Political Economy*, XI (March, 1903), 237–56.

Deutsch, Herman J. "Disintegrating Forces in Wisconsin Politics of the Early Seventies." *Wisconsin Magazine of History*, XV (December, March, June, 1931, and 1932), 168–81, 282–96. 391–411.

Dewey, Thelma A. "The National Grange Manual: 1875." *Michigan History*, XLVI (December, 1962), 330–32.

Easterby, J. H. "The Granger Movement in South Carolina." South Carolina Historical Association *Proceedings*, 1931, pp. 24–32.

Fairman, Charles. "The So-called Granger Cases, Lord Hales, and Justice Bradley." *Stanford Law Review*, V (July, 1953), 587–679.

"Farmer Folks Take a Holiday." *Minnesota History*, XL (Summer, 1967), 300–301.

Ferguson, James S. "Co-operative Activity of the Grange in Mississippi." *Journal of Mississippi History*, IV (January, 1942), 3–19.

————. "The Grange and Farmer Education in Mississippi." *Journal of Southern History*, VIII (November, 1942), 497–512.

Finneran, Helen T. "Records of the National Grange in its Washington Office." *The American Archivist*, XXVII (January, 1964), 103–11.

"Fitting Honor Paid to a Great Grange Pioneer; Memorial to Oliver Hudson Kelley." *National Grange Monthly*. XXXVI (January, 1939), 1, 22.

Fuller, Wayne E. "The Grange in Colorado." *The Colorado Magazine*, XXXVI (October, 1959), 254–65.

Galambos, Louis. "The Agrarian Image of the Large Corporation, 1879–1920: A Study in Social Accommodation." *Journal of Economic History*, XXVIII (September, 1968), 341–62.

Gates, Paul W. "Agricultural Change in New York State, 1850–1890." *New York History*, L (April, 1969), 115–41.

Glover, W. H. "The Agricultural College Crisis of 1885." *Wisconsin Magazine of History*, XXXII (September, 1948), 17–25.

"The Grange in the Great Valley." *Historical Review of Berks County*, XV (April, 1950), 194–97.

Hart, John F. "Loss and Abandonment of Cleared Farm Land in the Eastern United States." *Annals of the Association of American Geographers*, LVIII (September, 1968).

Hirsch, A. H. "Efforts of the Grange in the Middle West to Control the Price of Machinery." *Mississippi Valley Historical Review*, XV (March, 1929), 473–96.

"History of the Grange in Oregon as a Supplement Celebrating the Seventy-Fifth Anniversary of the State Grange." *Oregon Grange Bulletin* (October 5, 1948).

Holbrook, Stewart H. "Great Days of the Grangers." *American Mercury*, LXIII (August, 1946), 236–41.

Huyett, J. Burns. "Early Grange Activities in Jefferson County." *Magazine of the Jefferson County Historical Society*, V (December, 1939), 4–8.

Klement, Frank L. "Middle Western Copperheadism and the Genesis of the Granger Movement." *Mississippi Valley Historical Review*, XXXVIII (March, 1952), 679–94.

Lemmer, George F. "The Agricultural Program of a Leading Farm Periodical, *Colman's Rural World*." *Agricultural History*, XXIII (October, 1949), 245–53.

McCluggage, Robert. "Joseph H. Osborn, Grange Leader." *Wisconsin Magazine of History*, XXXV (Spring, 1952), 178–84.

Martin, Roscoe C. "The Grange as a Political Factor in Texas." *Southwestern Political and Social Science Quarterly*, VI (March, 1926), 363–83.

Mayhew, Anne. "A Reappraisal of the Causes of Farm Protest in the United States, 1870–1900." *Journal of Economic History*, XXXII (June, 1972), 464–75.

Merk, Frederick. "Eastern Antecedents of the Grangers." *Agricultural History*, XXIII (January, 1949), 1–8.

Metcalf, Henry H. "The New Hampshire State Grange." *Granite Monthly*, LIII (December, 1921), 517–26.

Miller, George H. "Chicago, Burlington and Quincy Railroad Company v. Iowa. *Iowa Journal of History*, LIV (October, 1956), 289–312.

————. "Origins of the Iowa Granger Law." *Mississippi Valley Historical Review*, XL (March, 1954), 657–80.

Millet, Donald J. "Some Aspects of Agricultural Retardation in Southwest

Louisiana, 1865–1900," *Louisiana History*, XI (Winter, 1970), 37–61.

Miner, H. Craig. "Hopes and Fears: Ambivalence in the Anti-Railroad Movement at Springfield, Missouri, 1870–1880," *Bulletin of the Missouri Historical Society*, XXVII (January, 1971), 129–46.

Mitchell, Humphrey. "The Grange in Canada." Queen's University, Canada, Departments of History, Political Science, and Economics, *Bulletin*, 13, 1911, pp. 1–20.

Mott, David C. "William Duane Wilson." *Annals of Iowa*, XX (July, 1936), 361–74.

Nash, Gerald D. "Origins of the Interstate Commerce Act of 1887." *Pennsylvania History*, XXIV (July, 1957), 181–90.

Noblin, Stuart. "Leonidas Lafayette Polk and the North Carolina Department of Agriculture." *North Carolina Historical Review*, XX (April, July, 1943), 103–21, 197–218.

Nordin, Dennis S. "A Revisionist Interpretation of the Patrons of Husbandry, 1867–1900," *Historian*, XXXII (August, 1970), 630–43.

———. "Graduate Studies in American Agricultural History." *Agricultural History*, XLI (July, 1967), 275–312.

Partin, Robert. "Black's Bend Grange, 1873–77: A Case Study of a Subordinate Grange of the Deep South." *Agricultural History*, XXXI (July, 1957), 49–59.

Paul, Rodman W. "The Great California Grain War: The Grangers Challenge the Wheat King." *Pacific Historical Review*, XXVII (November, 1958), 331–49.

Pillar, James J. "Catholic Opposition to the Grange in Mississippi," *Journal of Mississippi History*, XXXI (August, 1969), 215–28.

Pollack, Norman. "Hofstadter on Populism: A Critique of 'The Age of Reform.' " *Journal of Southern History*, XXVI (November, 1960), 478–500.

———. "The Myth of Populist Anti-Semitism." *American Historical Review*, LXVIII (October, 1962), 76–80.

Prescott, Gerald. "Wisconsin Farm Leaders in the Gilded Age." *Agricultural History*, XLIV (April, 1970), 183–99.

Reynold, Elmer E. "Turning Back History's Pages on a Half Century of the Grange." *National Grange Monthly*, XXXVI (November, 1939), 5, 14, 24.

Ridge, Martin. "Ignatius Donnelly and the Granger Movement in Minnesota." *Mississippi Valley Historical Review*, XLII (March, 1956), 693–709.

Rogers, James H. "The Grange—then and now." *Southern Planter*, XCVII (August, 1936), 6, 25.

Rogers, William W. "The Alabama State Fair, 1865–1900," *Alabama Review*, XI (April, 1958), 100–16.

———. "The Alabama State Grange." *Alabama Review*, VIII (April, 1955), 104–18.

Rosenberg, Charles E. "Science, Technology, and Economic Growth: The Case of the Agricultural Experiment Station Scientist, 1875–1914," *Agricultural History*, XLV (January, 1971), 1–20.

Ross, Earle D. "The Land-Grant College: A Democratic Adaptation." *Agricultural History*, XV (January, 1941), 26–36.

Rothstein, Morton. "America in the International Rivalry for the British

Wheat Market, 1860–1914," *Mississippi Valley Historical Review*, XLVII (December, 1960), 401–18.

Saloutos, Theodore. "The Grange in the South." *Journal of Southern History*, XIX (November, 1953), 473–87.

_____. "The Professors and the Populists." *Agricultural History*, XL (October, 1966), 235–54.

Schell, Herbert S. "The Grange and the Credit Problem in Dakota Territory." *Agricultural History*, X (April, 1936), 59–83.

Schmidt, Louis B. "Farm Organizations in Iowa." *Palimpsest*, XXXI (April, 1950), 117–64.

_____. "The History of the Granger Movement." *Prairie Farmer*. XCIII (January, 22, 29, February 5, 12, 19, 1921).

Scott, Roy V. "Grangerism in Champaign County, Illinois, 1873–1877." *Mid-America*, XLIII (July, 1961), 139–63.

Sherman, Rexford B. "The New Hampshire Grange, 1873–1883," *Historical New Hampshire*, XXVI (Spring, 1971), 2–25.

Sigmund, Edwin W. "The Granger Cases: 1877 or 1876?." *American Historical Review*, LVIII (April, 1953), 571–74.

Sinclair, Elsie C. "The Story of Consumer Co-operation in Indiana." *Indiana Magazine of History*, XXXIII (March, 1937), 45–51.

Smith, Ralph A. "The Contribution of the Grangers to Education in Texas." *Southwestern Social Science Quarterly*, XXI (March, 1941), 312–24.

_____. "The Co-operative Movement in Texas, 1870–1900." *Southwestern Historical Quarterly*, XLII (April, 1939), 297–315.

Taylor, Henry C. "I Knew Oliver Hudson Kelley." *Grange*, LXI (November, 1967), 7.

Throne, Mildred. "The Anti-Monopoly Party in Iowa, 1873–1874." *Iowa Journal of History*, LII (October, 1954), 289–326.

_____. "The Grange in Iowa, 1868–1875." *Iowa Journal of History*, XLVII (October, 1949), 289–324.

_____. "The Repeal of the Iowa Granger Law, 1878." *Iowa Journal of History*, LI (April, 1953), 97–130.

Tontz, Robert L. "Memberships of General Farmers' Organizations, United States, 1874–1960," *Agricultural History*, XXXVIII (July, 1964), 143–56.

Treleven, Dale E. "Railroads, Elevators, and Grain Dealers: The Genesis of Antimonopolism in Milwaukee." *Wisconsin Magazine of History*, LII (Spring, 1969), 205–22.

White, J. M. "Origin and Location of the Mississippi A. & M. College." *Mississippi Historical Society, Publications*, III (1900), 341–51.

Woodman, Harold D. "Chicago Businessmen and the 'Granger' Laws," *Agricultural History*, XXXVI (January, 1962), 16–24.

C. THESES AND DISSERTATIONS

Aldous, Lois G. "The Grange in Kansas since 1895," M.A. thesis, University of Kansas, 1941.

Armstrong, Lindsey O. "The Development of Agricultural Education in

North Carolina." M.S. thesis, North Carolina State College of Agriculture and Engineering, 1932.

Barns, William D. "The Granger and Populist Movements in West Virginia, 1873–1914." Ph.D. dissertation, West Virginia University, 1946.

Barton, Richard H. "The Agrarian Revolt in Michigan, 1865–1900." Ph.D. dissertation, Michigan State University, 1958.

Beinhauer, Myrtle T. "History of Farm Organizations in Iowa, 1838–1931." M.A. thesis, Drake University, 1932.

Bittinger, Richard D. "History of the Patrons of Husbandry in Kansas, 1872–1882," M.A. thesis, University of Kansas, 1960.

Broadway, H. H. "Frank Burkitt: The Man in the Wool Hat." M.A. thesis, Mississippi State College, 1948.

Carter, Floella K. "The Grange in Missouri, 1878–1939." M.A. thesis, University of Missouri, 1940.

Chapman, Harry A. "The Historical Development of the Grange in South Carolina." M.A. thesis, Furman University, 1951.

Clevenger, Homer. "Agrarian Politics in Missouri, 1880–1896." Ph.D. dissertation, University of Missouri, 1940.

Cole, Houston. "Populism in Tuscaloosa." M.A. thesis, University of Alabama, 1927.

Dailey, Alan D. "Agrarian Movements in Indiana, 1870–1896, Compared with the Movements in Other Middle Western States." M.A. thesis, Indiana University, 1947.

Davis, Granville D. "The Granger Movement in Arkansas." M.A. thesis, University of Illinois, 1931.

Evans, Samuel L. "Texas Agriculture, 1880–1930." Ph.D. dissertation, University of Texas, 1960.

Ferguson, James S. "Agrarianism in Mississippi, 1871–1900—A Study in Nonconformity." Ph.D. dissertation, University of North Carolina, 1952.

Fuller, Wayne E. "History of the Grange in Colorado." M.A. thesis, University of Denver, 1948.

———. "R.F.D.: A History of the Farmers' Mail." Ph.D. dissertation, University of California, Berkeley, 1954.

Hair, William I. "The Agrarian Protest in Louisiana, 1877–1900." Ph.D. dissertation, Louisiana State University, 1962.

Hebb, Douglas. "The Woman Movement in the California State Grange." M.A. thesis, University of California, Berkeley, 1950.

Johnson, Carleton Ware. "Social Life of Central Illinois in the 1870's." M.A. thesis, University of Illinois, 1933.

Keppel, Ann M. "Country Schools for Country Children: Backgrounds of the Reform Movement in Rural Elementary Education, 1890–1914." Ph.D. dissertation, University of Wisconsin, 1960.

Kretschmann, James F. "The North Carolina Department of Agriculture, 1877–1900." M.A. thesis, University of North Carolina, 1955.

Lea, Charles W. "The Grange Movement in Wisconsin." M.A. thesis, University of Wisconsin, 1897.

Legg, Bryon S. "The Granger Movement in Indiana, 1869–1880." M.A. thesis, Indiana University, 1915.

Leonard, Sister Mary T. C. "The Granger Movement in Specific Relation to Nebraska." M.A. thesis, De Paul University, 1950.

Lever, Webbie J. "The Agrarian Movement in Noxubee County." M.A. thesis, Mississippi State College, 1952.

McDaniel, Curtis E. "Educational and Social Interests of the Grange in Texas, 1873–1905." M.A. thesis, University of Texas, 1938.

McGuire, Peter S. "The Genesis of the Interstate Commerce Commission." Ph.D. dissertation, Cornell University, 1923.

Miller, George. "The Granger Laws: A Study of the Origins of State Railway Control in the Upper Mississippi Valley." Ph.D. dissertation, University of Michigan, 1951.

Murphy, John E. "Recreational Facilities of the Midwestern Farmers, 1860–1900." M.A. thesis, University of Illinois, 1943.

Nydahl, Theodore L. "The Diary of Ignatius Donnelly, 1859–1884." Ph.D. dissertation, University of Minnesota, 1942.

Oostensorp, John A. "The Cooperative Movement in the Patrons of Husbandry, in Iowa, 1870–1878." M.A. thesis, University of Iowa, 1949.

Owsley, Carol L. "The History of Early Agricultural Societies in Kansas." M.S. thesis, Kansas State College, 1947.

Prescott, Gerald L. "Yeomen, Entrepreneurs and Gentry: A Comparative Study of Three Wisconsin Agricultural Organizations, 1873–1893." Ph.D. dissertation, University of Wisconsin, 1969.

Rogers, William W. "Agrarianism in Alabama, 1865–1896." Ph.D. dissertation, University of North Carolina, 1959.

Scott, Edna A. "The Grange Movement in Oregon, 1873–1900." M.A. thesis, University of Oregon, 1923.

Sherman, Rexford B. "The Grange in Maine and New Hampshire, 1870–1940." Ph.D. dissertation, Boston University, 1973.

Shirey, Mervin R. "The Granger Movement in West Virginia." M.A. thesis, West Virginia University, 1933.

Shoemaker, Rose M. "The Granger Movement in Iowa." M.A. thesis, University of Chicago, 1927.

Skrepetos, Venetta. "A Study of the California State Grange as a Pressure Group in California." M.S. thesis, University of California, Berkeley, 1958.

Smith, J. Harold. "History of the Grange in Kansas, 1883–1897." M.A. thesis, University of Kansas, 1940.

Smith, O. P. "Farm Organizations in Alabama from 1872 to 1907." M.A. thesis, Auburn University, 1940.

Smith, Ralph A. "A. J. Rose, Agrarian Crusader of Texas." Ph.D. dissertation, University of Texas, 1938.

Smith, R. E. "Wisconsin Granger Movement." M.A. thesis, University of Wisconsin, 1895.

Stephenson, Sarah M. "The Social and Educational Aspects of the Grange, 1870–1934." M.A. thesis, University of Wisconsin, 1935.

Throne, Mildred. "A History of Agriculture in Southern Iowa, 1833–1890." Ph.D. dissertation, University of Iowa, 1946.

Wentworth, Bertha E. "The Influence of the Grange Movement upon the

Educational and Social Development of the Agricultural Class of Kansas from 1872–1876." M.S. thesis, Kansas State College, 1929.

Wilkinson, Alma B. "The Granger Movement in Missouri." M.A. thesis, University of Missouri, 1926.

Willis, Curley D. "The Grange Movement in Louisiana." M.A. thesis, Louisiana State University, 1945.

Wright, John I. "Enforcement of the Granger Laws in Illinois, 1870–1886." M.A. thesis, University of Chicago, 1934.

RICH HARVEST

A History of the Grange
1867-1900

D. SVEN NORDIN

The **Rich Harvest** tells of the development and progress of the Order of Patrons of Husbandry, beginning with Oliver Hudson Kelley's first activities on behalf of the farmer organization. It represents the first scholarly work devoted to the history of the Grange.

Oliver Kelley concluded that isolation and ignorance were the greatest rural problems confronting the nation and that something could be done to eliminate them. As a result, he created a national farm organization that, by working through subordinate chapters and with the Department of Agriculture, could help to eliminate many of these shortcomings. The new agrarian brotherhood was named the Order of Patrons of Husbandry or simply, the Grange. The original educational and social objectives of the society were subsequently expanded to include other reforms, but Kelley's first goals for the Grange remained the most significant of the order's activities.

The study under consideration represents an attempt at revisionism. Emphasis has been given in the **Rich Harvest** to the Grange's social and educational objectives, not its economic and political aspects. Moreover, an explanation is given for the resiliency of the Grange in the eighties and nineties. What emerges then is an account which balances out the Grange's endeavors and gives credence to the fact that the order remained true to its founders' original objectives for the organization. Coming

continued on back flap